Learn

Eureka Math®
Grade 3
Modules 1 & 2

Published by Great Minds®.

Copyright © 2018 Great Minds®.

Printed in the U.S.A.
This book may be purchased from the publisher at eureka-math.org.
BAB   10   9   8   7   6   5   4

ISBN 978-1-64054-060-6

G3-M1-M2-L-05.2018

Learn ◆ Practice ◆ Succeed

Eureka Math® student materials for *A Story of Units®* (K–5) are available in the *Learn, Practice, Succeed* trio. This series supports differentiation and remediation while keeping student materials organized and accessible. Educators will find that the *Learn, Practice,* and *Succeed* series also offers coherent—and therefore, more effective—resources for Response to Intervention (RTI), extra practice, and summer learning.

Learn

Eureka Math Learn serves as a student's in-class companion where they show their thinking, share what they know, and watch their knowledge build every day. *Learn* assembles the daily classwork—Application Problems, Exit Tickets, Problem Sets, templates—in an easily stored and navigated volume.

Practice

Each *Eureka Math* lesson begins with a series of energetic, joyous fluency activities, including those found in *Eureka Math Practice.* Students who are fluent in their math facts can master more material more deeply. With *Practice,* students build competence in newly acquired skills and reinforce previous learning in preparation for the next lesson.

Together, *Learn* and *Practice* provide all the print materials students will use for their core math instruction.

Succeed

Eureka Math Succeed enables students to work individually toward mastery. These additional problem sets align lesson by lesson with classroom instruction, making them ideal for use as homework or extra practice. Each problem set is accompanied by a Homework Helper, a set of worked examples that illustrate how to solve similar problems.

Teachers and tutors can use *Succeed* books from prior grade levels as curriculum-consistent tools for filling gaps in foundational knowledge. Students will thrive and progress more quickly as familiar models facilitate connections to their current grade-level content.

Students, families, and educators:

Thank you for being part of the *Eureka Math®* community, where we celebrate the joy, wonder, and thrill of mathematics.

In the *Eureka Math* classroom, new learning is activated through rich experiences and dialogue. The *Learn* book puts in each student's hands the prompts and problem sequences they need to express and consolidate their learning in class.

What is in the Learn book?

Application Problems: Problem solving in a real-world context is a daily part of *Eureka Math.* Students build confidence and perseverance as they apply their knowledge in new and varied situations. The curriculum encourages students to use the RDW process—Read the problem, Draw to make sense of the problem, and Write an equation and a solution. Teachers facilitate as students share their work and explain their solution strategies to one another.

Problem Sets: A carefully sequenced Problem Set provides an in-class opportunity for independent work, with multiple entry points for differentiation. Teachers can use the Preparation and Customization process to select "Must Do" problems for each student. Some students will complete more problems than others; what is important is that all students have a 10-minute period to immediately exercise what they've learned, with light support from their teacher.

Students bring the Problem Set with them to the culminating point of each lesson: the Student Debrief. Here, students reflect with their peers and their teacher, articulating and consolidating what they wondered, noticed, and learned that day.

Exit Tickets: Students show their teacher what they know through their work on the daily Exit Ticket. This check for understanding provides the teacher with valuable real-time evidence of the efficacy of that day's instruction, giving critical insight into where to focus next.

Templates: From time to time, the Application Problem, Problem Set, or other classroom activity requires that students have their own copy of a picture, reusable model, or data set. Each of these templates is provided with the first lesson that requires it.

Where can I learn more about Eureka Math resources?

The Great Minds® team is committed to supporting students, families, and educators with an ever-growing library of resources, available at eureka-math.org. The website also offers inspiring stories of success in the *Eureka Math* community. Share your insights and accomplishments with fellow users by becoming a *Eureka Math* Champion.

Best wishes for a year filled with aha moments!

Jill Diniz

Jill Diniz
Director of Mathematics
Great Minds

The Read–Draw–Write Process

The *Eureka Math* curriculum supports students as they problem-solve by using a simple, repeatable process introduced by the teacher. The Read–Draw–Write (RDW) process calls for students to

1. Read the problem.

2. Draw and label.

3. Write an equation.

4. Write a word sentence (statement).

Educators are encouraged to scaffold the process by interjecting questions such as

- What do you see?

- Can you draw something?

- What conclusions can you make from your drawing?

The more students participate in reasoning through problems with this systematic, open approach, the more they internalize the thought process and apply it instinctively for years to come.

Contents

Module 1: Properties of Multiplication and Division and Solving Problems with Units of 2–5 and 10

Module 2: Place Value and Problem Solving with Units of Measure

Grade 3
Module 1

There are 83 girls and 76 boys in the third grade. How many total students are in the third grade?

_____

_____

_____

_____

Read   **Draw**   **Write**

Name _____ Date _____

1. Fill in the blanks to make true statements.

a. 3 groups of five = _____

   3 fives = _____

   3 × 5 = _____

b. 3 + 3 + 3 + 3 + 3 = _____

   5 groups of three = _____

   5 × 3 = _____

c. 6 + 6 + 6 + 6 = _____

   _____ groups of six = _____

   4 × _____ = _____

d. 4 + ____ + ____ + ____ + ____ + ____ = _____

   6 groups of _____ = _____

   6 × _____ = _____

2. The picture below shows 2 groups of apples. Does the picture show 2 × 3? Explain why or why not.

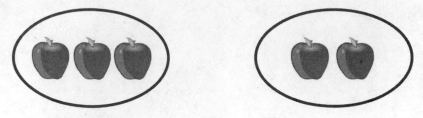

3. Draw a picture to show 2 × 3 = 6.

4. Caroline, Brian, and Marta share a box of chocolates. They each get the same amount. Circle the chocolates below to show 3 groups of 4. Then, write a repeated addition sentence and a multiplication sentence to represent the picture.

Lesson 1:   Understand *equal groups of* as multiplication.

Name _____   Date _____

1. The picture below shows 4 groups of 2 slices of watermelon. Fill in the blanks to make true repeated addition and multiplication sentences that represent the picture.

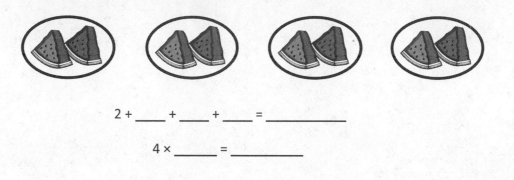

2 + _____ + _____ + _____ = _____

4 × _____ = _____

2. Draw a picture to show 3 + 3 + 3 = 9. Then, write a multiplication sentence to represent the picture.

Jordan uses 3 lemons to make 1 pitcher of lemonade. He makes 4 pitchers. How many lemons does he use altogether?

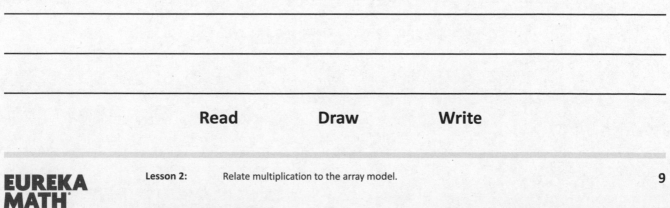

Read   **Draw**   **Write**

EUREKA
MATH®

Name _____   Date _____

Use the arrays below to answer each set of questions.

1.

a. How many rows of cars are there? _____

b. How many cars are there in each row? _____

2.

a. What is the number of rows? _____

b. What is the number of objects in each row? _____

3.

a. There are 4 spoons in each row. How many spoons are in 2 rows? _____

b. Write a multiplication expression to describe the array. _____

4.

a. There are 5 rows of triangles. How many triangles are in each row? _____

b. Write a multiplication expression to describe the total number of triangles.

_____

5. The dots below show 2 groups of 5.

       a. Redraw the dots as an array that shows 2 rows of 5.

       b. Compare the drawing to your array. Write at least 1 reason why they are the same and 1 reason why they are different.

6. Emma collects rocks. She arranges them in 4 rows of 3. Draw Emma's array to show how many rocks she has altogether. Then, write a multiplication equation to describe the array.

7. Joshua organizes cans of food into an array. He thinks, "My cans show 5 × 3!" Draw Joshua's array to find the total number of cans he organizes.

EUREKA
MATH

Name _____   Date _____

1.   ★ ★ ★
     ★ ★ ★
     ★ ★ ★
     ★ ★ ★

a. There are 4 rows of stars. How many stars are in each row? _____

b. Write a multiplication equation to describe the array. _____

2. Judy collects seashells. She arranges them in 3 rows of 6. Draw Judy's array to show how many seashells she has altogether. Then, write a multiplication equation to describe the array.

_____
threes array

Robbie sees that a carton of eggs shows an array with 2 rows of 6 eggs. What is the total number of eggs in the carton?

_____

_____

_____

_____

Read   **Draw**   **Write**

EUREKA
MATH·

Lesson 3:   Interpret the meaning of factors—the size of the group or the number
             of groups.

17

© 2018 Great Minds®. eureka-math.org

Name _____   Date _____

Solve Problems 1–4 using the pictures provided for each problem.

1. There are 5 flowers in each bunch. How many flowers are in 4 bunches?

       a. Number of groups: _____   Size of each group: _____

       b. $4 \times 5 =$ _____

       c. There are _____ flowers altogether.

2. There are _____ candies in each box. How many candies are in 6 boxes?

       a. Number of groups: _____   Size of each group: _____

       b. $6 \times$ _____ = _____

       c. There are _____ candies altogether.

3. There are 4 oranges in each row. How many oranges are there in _____ rows?

       a. Number of rows: _____   Size of each row: _____

       b. _____ $\times 4 =$ _____

       c. There are _____ oranges altogether.

Lesson 3:   Interpret the meaning of factors—the size of the group or the number of groups.

4. There are _____ loaves of bread in each row. How many loaves of bread are there in 5 rows?

a. Number of rows: _____   Size of each row: _____

b. _____ × _____ = _____

c. There are _____ loaves of bread altogether.

5. a. Write a multiplication equation for the array shown below.

X X X

X X X

X X X

X X X

b. Draw a number bond for the array where each part represents the amount in one row.

6. Draw an array using factors 2 and 3. Then, show a number bond where each part represents the amount in one row.

Lesson 3:   Interpret the meaning of factors—the size of the group or the number of groups.

© 2018 Great Minds®. eureka-math.org

Name _____   Date _____

Draw an array that shows 5 rows of 3 squares. Then, show a number bond where each part represents the amount in one row.

EUREKA MATH

Lesson 3:   Interpret the meaning of factors—the size of the group or the number of groups.

© 2018 Great Minds®. eureka-math.org

21

The student council holds a meeting in Mr. Chang's classroom. They arrange the chairs in 3 rows of 5. How many chairs are used in all?

_____

_____

_____

_____

Read   **Draw**   **Write**

Name _____   Date _____

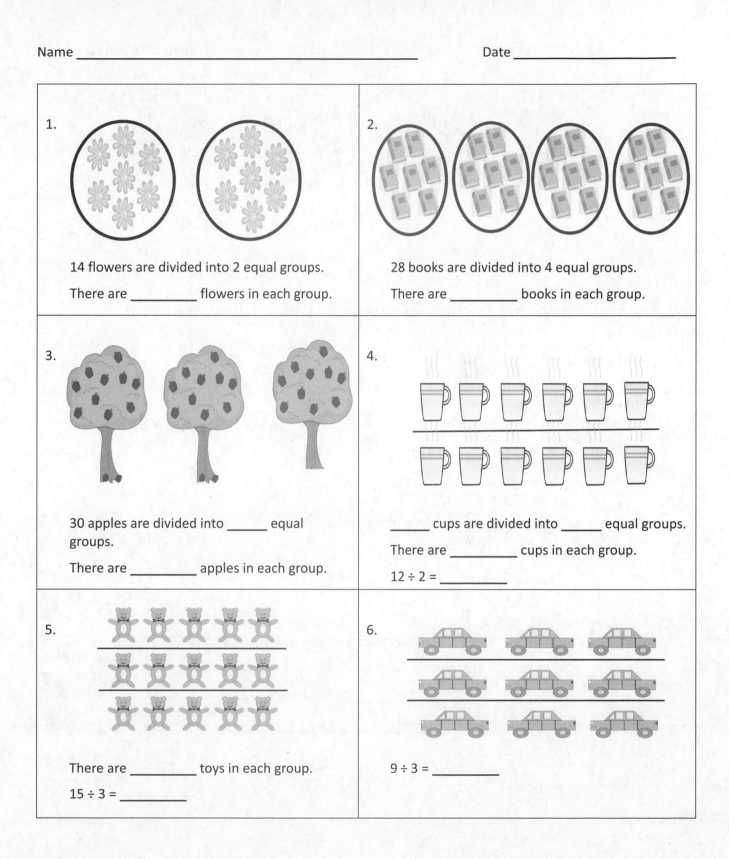

1. 14 flowers are divided into 2 equal groups.

There are _____ flowers in each group.

2. 28 books are divided into 4 equal groups.

There are _____ books in each group.

3. 30 apples are divided into _____ equal groups.

There are _____ apples in each group.

4. _____ cups are divided into _____ equal groups.

There are _____ cups in each group.

$12 \div 2 =$ _____

5. There are _____ toys in each group.

$15 \div 3 =$ _____

6. $9 \div 3 =$ _____

Lesson 4:   Understand the meaning of the unknown as the size of the group in division.

25

EUREKA MATH®

7. Audrina has 24 colored pencils. She puts them in 4 equal groups. How many colored pencils are in each group?

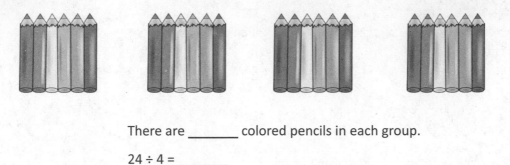

There are _____ colored pencils in each group.

24 ÷ 4 = _____

8. Charlie picks 20 apples. He divides them equally between 5 baskets. Draw the apples in each basket.

There are _____ apples in each basket.

20 ÷ _____ = _____

9. Chelsea collects butterfly stickers. The picture shows how she placed them in her book. Write a division sentence to show how she equally grouped her stickers.

There are _____ butterflies in each row.

_____ ÷ _____ = _____

Lesson 4:   Understand the meaning of the unknown as the size of the group in division.

Name _____ Date _____

1. There are 16 glue sticks for the class. The teacher divides them into 4 equal groups. Draw the number of glue sticks in each group.

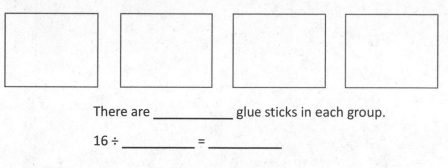

There are _____ glue sticks in each group.

16 ÷ _____ = _____

2. Draw a picture to show 15 ÷ 3. Then, fill in the blank to make a true division sentence.

15 ÷ 3 = _____

Lesson 4:   Understand the meaning of the unknown as the size of the group in division.

27

© 2018 Great Minds®. eureka-math.org

Stacey has 18 bracelets. After she organizes the bracelets by color, she has 3 equal groups. How many bracelets are in each group?

_____

_____

_____

_____

Read   **Draw**   **Write**

Lesson 5:   Understand the meaning of the unknown as the number of groups in division.

© 2018 Great Minds®. eureka-math.org

29

Name _____   Date _____

1.

Divide 6 tomatoes into groups of 3.

There are _____ groups of 3 tomatoes.

6 ÷ 3 = 2

2.

Divide 8 lollipops into groups of 2.

There are _____ groups.

8 ÷ 2 = _____

3.

Divide 10 stars into groups of 5.

10 ÷ 5 = _____

4.

Divide the shells to show 12 ÷ 3 = _____, where the unknown represents the number of groups.

How many groups are there? _____

5. Rachel has 9 crackers. She puts 3 crackers in each bag. Circle the crackers to show Rachel's bags.

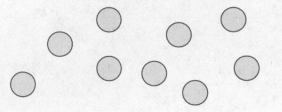

a. Write a division sentence where the answer represents the number of Rachel's bags.

b. Draw a number bond to represent the problem.

6. Jameisha has 16 wheels to make toy cars. She uses 4 wheels for each car.

a. Use a count-by to find the number of cars Jameisha can build. Make a drawing to match your counting.

b. Write a division sentence to represent the problem.

Lesson 5:   Understand the meaning of the unknown as the number of groups in division.

© 2018 Great Minds®. eureka-math.org

Name _____   Date _____

1. Divide 12 triangles into groups of 6.

12 ÷ 6 = _____

2. Spencer buys 20 strawberries to make smoothies. Each smoothie needs 5 strawberries. Use a count-by to find the number of smoothies Spencer can make. Make a drawing to match your counting.

Lesson 5:   Understand the meaning of the unknown as the number of groups in division.

© 2018 Great Minds®. eureka-math.org

33

Twenty children play a game. There are 5 children on each team. How many teams play the game?

Write a division sentence to represent the problem.

_____

_____

_____

_____

Read   **Draw**   **Write**

EUREKA
MATH®

Name _____ Date _____

1. Rick puts 15 tennis balls into cans. Each can holds 3 balls. Circle groups of 3 to show the balls in each can.

Rick needs _____ cans.

_____ × 3 = 15

15 ÷ 3 = _____

2. Rick uses 15 tennis balls to make 5 equal groups. Draw to show how many tennis balls are in each group.

There are _____ tennis balls in each group.

5 × _____ = 15

15 ÷ 5 = _____

3. Use an array to model Problem 1.

a. _____ × 3 = 15

    15 ÷ 3 = _____

    The number in the blanks represents

    _____.

b. 5 × _____ = 15

    15 ÷ 5 = _____

    The number in the blanks represents

    _____.

4. Deena makes 21 jars of tomato sauce. She puts 7 jars in each box to sell at the market. How many boxes does Deena need?

    21 ÷ 7 = _____

    _____ × 7 = 21

    What is the meaning of the unknown factor and quotient? _____

    _____

5. The teacher gives the equation 4 × _____ = 12. Charlie finds the answer by writing and solving
    12 ÷ 4 = _____. Explain why Charlie's method works.

    _____

6. The blanks in Problem 5 represent the size of the groups. Draw an array to represent the equations.

Lesson 6:   Interpret the unknown in division using the array model.

EUREKA
MATH

Name _____ Date _____

Cesar arranges 12 notecards into rows of 6 for his presentation. Draw an array to represent the problem.

$12 \div 6 =$ _____

_____ $\times 6 = 12$

What do the unknown factor and quotient represent? _____

Anna picks 24 flowers. She makes equal bundles of flowers and gives 1 bundle to each of her 7 friends. She keeps a bundle for herself too. How many flowers does Anna put in each bundle?

_____

_____

_____

_____

Read   Draw   Write

EUREKA
MATH

Lesson 7:   Demonstrate the commutativity of multiplication, and practice related
              facts by skip-counting objects in array models.

© 2018 Great Minds®. eureka-math.org

41

Name _____   Date _____

1. a. Draw an array that shows 6 rows of 2.

    b. Write a multiplication sentence where the first factor represents the number of rows.

        _____ × _____ = _____

2. a. Draw an array that shows 2 rows of 6.

    b. Write a multiplication sentence where the first factor represents the number of rows.

        _____ × _____ = _____

3. a. Turn your paper to look at the arrays in Problems 1 and 2 in different ways. What is the same and what is different about them?

    b. Why are the factors in your multiplication sentences in a different order?

4. Write a multiplication sentence for each expression. You might skip-count to find the totals.

    a. 6 twos: ___6 × 2 = 12___   d. 2 sevens: _____   **Extension:**

    b. 2 sixes: _____   e. 9 twos: _____   g. 11 twos: _____

    c. 7 twos: _____   f. 2 nines: _____   h. 2 twelves: _____

Lesson 7:   Demonstrate the commutativity of multiplication, and practice related facts by skip-counting objects in array models.

43

EUREKA
MATH

5. Write and solve multiplication sentences where the second factor represents the size of the row.

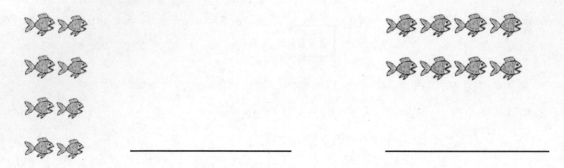

_____   _____

6. Ms. Nenadal writes 2 × 7 = 7 × 2 on the board. Do you agree or disagree? Draw arrays to help explain your thinking.

7. Find the missing factor to make each equation true.

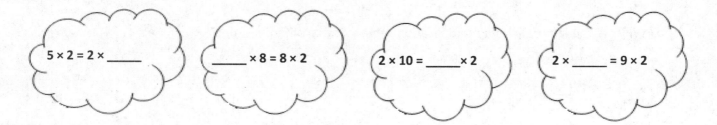

5 × 2 = 2 × _____   _____ × 8 = 8 × 2   2 × 10 = _____ × 2   2 × _____ = 9 × 2

8. Jada gets 2 new packs of erasers. Each pack has 6 erasers in it.

    a. Draw an array to show how many erasers Jada has altogether.

    b. Write and solve a multiplication sentence to describe the array.

    c. Use the commutative property to write and solve a different multiplication sentence for the array.

Lesson 7:   Demonstrate the commutativity of multiplication, and practice related
            facts by skip-counting objects in array models.

© 2018 Great Minds®. eureka-math.org

Name _____   Date _____

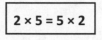

$$2 \times 5 = 5 \times 2$$

Do you agree or disagree with the statement in the box? Draw arrays and use skip-counting to explain your thinking.

Lesson 7:   Demonstrate the commutativity of multiplication, and practice related   45
               facts by skip-counting objects in array models.

© 2018 Great Minds®. eureka-math.org

Children sit in 2 rows of 9 on the carpet for math time. Erin says, "We make 2 equal groups." Vittesh says, "We make 9 equal groups." Who is correct? Explain how you know using models, numbers, and words.

_____

_____

_____

_____

Read   **Draw**   **Write**

EUREKA MATH

Lesson 8:   Demonstrate the commutativity of multiplication, and practice related facts by skip-counting objects in array models.

47

Name _____   Date _____

1. Draw an array that shows 5 rows of 3.

2. Draw an array that shows 3 rows of 5.

3. Write multiplication expressions for the arrays in Problems 1 and 2. Let the first factor in each expression represent the number of rows. Use the commutative property to make sure the equation below is true.

_____ × _____ = _____ × _____
   Problem 1   **Problem 2**

4. Write a multiplication sentence for each expression. You might skip-count to find the totals. The first one is done for you.

a. 2 threes: _____$2 \times 3 = 6$_____

d. 4 threes: _____

g. 3 nines: _____

b. 3 twos: _____

e. 3 sevens: _____

h. 9 threes: _____

c. 3 fours: _____

f. 7 threes: _____

i. 10 threes: _____

5. Find the unknowns that make the equations true. Then, draw a line to match related facts.

a. 3 + 3 + 3 + 3 + 3 = _____

d. $3 \times 8 =$ _____

b. $3 \times 9 =$ _____

e. _____ $= 5 \times 3$

c. 7 threes + 1 three = _____

f. $27 = 9 \times$ _____

EUREKA MATH

Lesson 8: Demonstrate the commutativity of multiplication, and practice related facts by skip-counting objects in array models.

© 2018 Great Minds®. eureka-math.org

49

6. Isaac picks 3 tangerines from his tree every day for 7 days.

   a. Use circles to draw an array that represents the tangerines Isaac picks.

   b. How many tangerines does Isaac pick in 7 days? Write and solve a multiplication sentence to find the total.

   c. Isaac decides to pick 3 tangerines every day for 3 more days. Draw x's to show the new tangerines on the array in Part (a).

   d. Write and solve a multiplication sentence to find the total number of tangerines Isaac picks.

7. Sarah buys bottles of soap. Each bottle costs $2.

   a. How much money does Sarah spend if she buys 3 bottles of soap?

   _____ × _____ = $_____

   b. How much money does Sarah spend if she buys 6 bottles of soap?

   _____ × _____ = $_____

Lesson 8:   Demonstrate the commutativity of multiplication, and practice related facts by skip-counting objects in array models.

Name _____   Date _____

Mary Beth organizes stickers on a page in her sticker book. She arranges them in 3 rows and 4 columns.

a. Draw an array to show Mary Beth's stickers.

b. Use your array to write a multiplication sentence to find Mary Beth's total number of stickers.

c. Label your array to show how you skip-count to solve your multiplication sentence.

d. Use what you know about the commutative property to write a different multiplication sentence for your array.

Lesson 8:   Demonstrate the commutativity of multiplication, and practice related
              facts by skip-counting objects in array models.

© 2018 Great Minds®. eureka-math.org

51

Name _____   Date _____

1. The team organizes soccer balls into 2 rows of 5. The coach adds 3 rows of 5 soccer balls. Complete the equations to describe the total array.

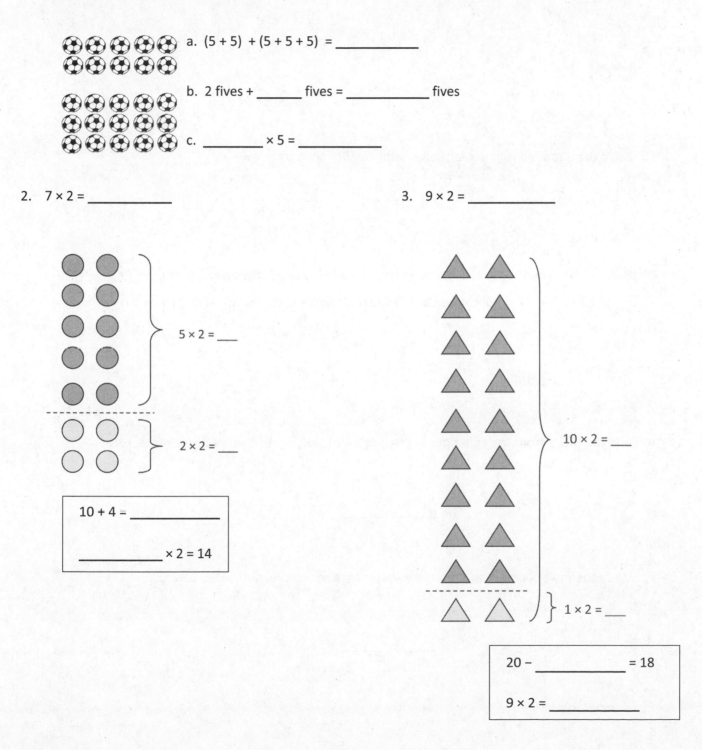

a. (5 + 5) + (5 + 5 + 5) = _____

b. 2 fives + _____ fives = _____ fives

c. _____ × 5 = _____

2. 7 × 2 = _____

5 × 2 = ___

2 × 2 = ___

10 + 4 = _____

_____ × 2 = 14

3. 9 × 2 = _____

10 × 2 = ___

1 × 2 = ___

20 − _____ = 18

9 × 2 = _____

EUREKA MATH

Lesson 9:   Find related multiplication facts by adding and subtracting equal groups in array models.

53

© 2018 Great Minds®. eureka-math.org

4. Matthew organizes his baseball cards in 4 rows of 3.

     a. Draw an array that represents Matthew's cards using an x to show each card.

     b. Solve the equation to find Matthew's total number of cards. $4 \times 3 =$ _____

5. Matthew adds 2 more rows. Use circles to show his new cards on the array in Problem 4(a).

     a. Write and solve a multiplication equation to represent the circles you added to the array.

                     _____ $\times 3 =$ _____

     b. Add the totals from the equations in Problems 4(b) and 5(a) to find Matthew's total cards.

                     _____ $+$ _____ $= 18$

     c. Write the multiplication equation that shows Matthew's total number of cards.

                     _____ $\times$ _____ $= 18$

Name _____   Date _____

1. Mrs. Stern roasts cloves of garlic. She places 10 rows of two cloves on a baking sheet.

    Write an equation to describe the number of cloves Mrs. Stern bakes.

    _____ × _____ = _____

2. When the garlic is roasted, Mrs. Stern uses some for a recipe. There are 2 rows of two garlic cloves left on the pan.

    a. Complete the equation below to show how many garlic cloves Mrs. Stern uses.

        _____ twos – _____ twos = _____ twos

    b. 20 – _____ = 16

    c. Write an equation to describe the number of garlic cloves Mrs. Stern uses.

        _____ × 2 = _____

EUREKA
MATH®

Lesson 9:   Find related multiplication facts by adding and subtracting equal
             groups in array models.

© 2018 Great Minds®. eureka-math.org

55

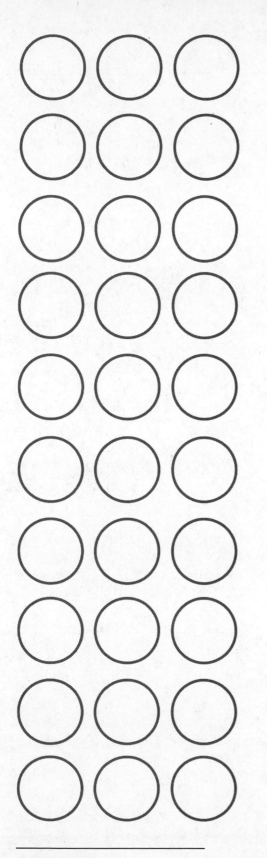

threes array no fill

EUREKA MATH

Lesson 9: Find related multiplication facts by adding and subtracting equal groups in array models.

57

© 2018 Great Minds®. eureka-math.org

A guitar has 6 strings. How many strings are there on 3 guitars? Write a multiplication equation to solve.

_____

_____

_____

_____

Read   **Draw**   **Write**

EUREKA MATH

Lesson 10:   Model the distributive property with arrays to decompose units as a strategy to multiply.

© 2018 Great Minds®. eureka-math.org

59

Name _____   Date _____

1. $7 \times 3 = (5 \times 3) + (2 \times 3) = $ _____

2. $8 \times 3 = (4 \times 3) + (4 \times 3) = $ _____

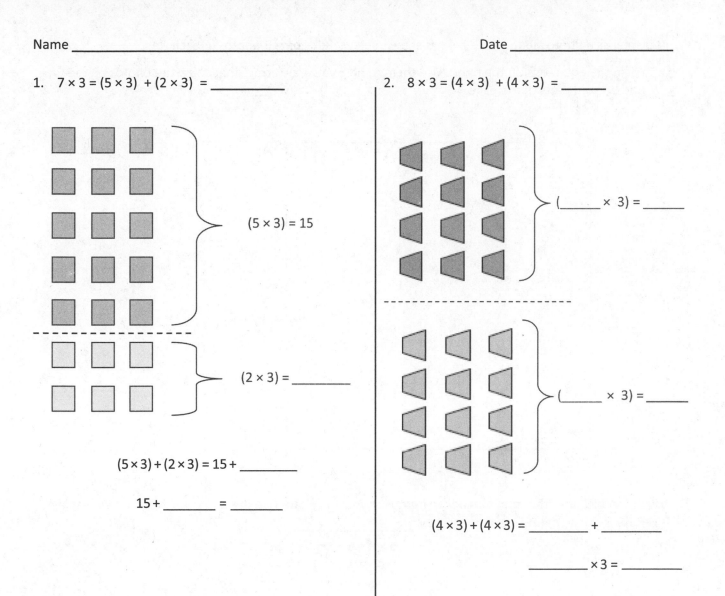

$(5 \times 3) = 15$

$(2 \times 3) = $ _____

$(5 \times 3) + (2 \times 3) = 15 + $ _____

$15 + $ _____ $ = $ _____

$(\underline{\hspace{1cm}} \times 3) = $ _____

$(\underline{\hspace{1cm}} \times 3) = $ _____

$(4 \times 3) + (4 \times 3) = $ _____ $ + $ _____

_____ $ \times 3 = $ _____

Lesson 10:   Model the distributive property with arrays to decompose units as a
                  strategy to multiply.

© 2018 Great Minds®. eureka-math.org

61

3. Ruby makes a photo album. One page is shown below. Ruby puts 3 photos in each row.

   a. Fill in the equations on the right. Use them to help you draw arrays that show the photos on the top and bottom parts of the page.

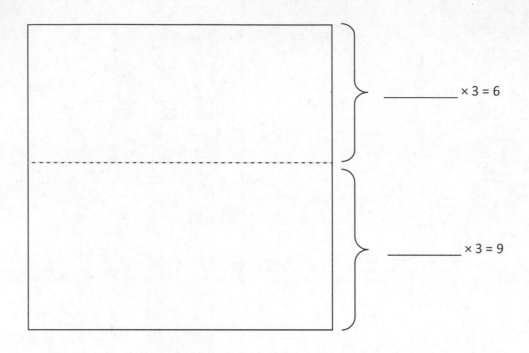

_____ × 3 = 6

_____ × 3 = 9

   b. Ruby calculates the total number of photos as shown below. Use the array you drew to help explain Ruby's calculation.

$$5 \times 3 = 6 + 9 = 15$$

Lesson 10:   Model the distributive property with arrays to decompose units as a strategy to multiply.

© 2018 Great Minds®. eureka-math.org

Name _____   Date _____

1. 6 × 3 = _____

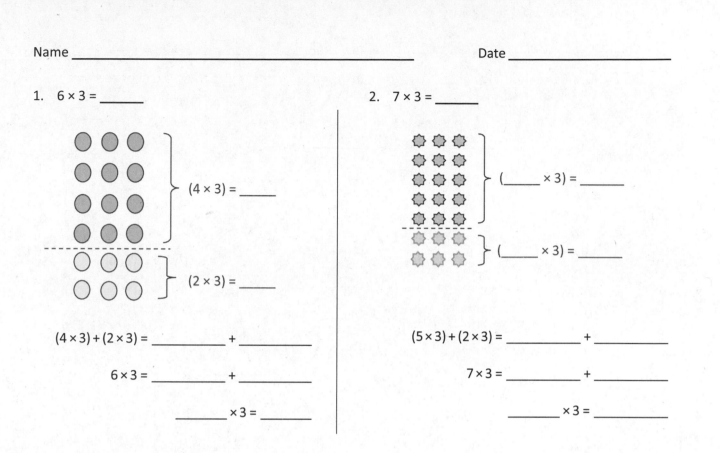

(4 × 3) = _____

(2 × 3) = _____

(4 × 3) + (2 × 3) = _____ + _____

6 × 3 = _____ + _____

_____ × 3 = _____

2. 7 × 3 = _____

(_____ × 3) = _____

(_____ × 3) = _____

(5 × 3) + (2 × 3) = _____ + _____

7 × 3 = _____ + _____

_____ × 3 = _____

Lesson 10:   Model the distributive property with arrays to decompose units as a strategy to multiply.

© 2018 Great Minds®. eureka-math.org

63

Rosie puts 2 lemon slices in each cup of iced tea. She uses a total of 8 slices. How many cups of iced tea does Rosie make?

_____

_____

_____

Read   **Draw**   **Write**

EUREKA
MATH®

Lesson 11:   Model division as the unknown factor in multiplication using arrays
               and tape diagrams.

65

© 2018 Great Minds®. eureka-math.org

Name _____   Date _____

1. Mrs. Prescott has 12 oranges. She puts 2 oranges in each bag. How many bags does she have?

    a. Draw an array where each column shows a bag of oranges.

    _____ ÷ 2 = _____

    b. Redraw the oranges in each bag as a unit in the tape diagram. The first unit is done for you. As you draw, label the diagram with known and unknown information from the problem.

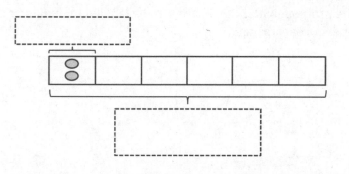

2. Mrs. Prescott arranges 18 plums into 6 bags. How many plums are in each bag? Model the problem with both an array and a labeled tape diagram. Show each column as the number of plums in each bag.

    There are _____ plums in each bag.

Lesson 11:   Model division as the unknown factor in multiplication using arrays and tape diagrams.

© 2018 Great Minds®. eureka-math.org

67

3. Fourteen shopping baskets are stacked equally in 7 piles. How many baskets are in each pile? Model the problem with both an array and a labeled tape diagram. Show each column as the number of baskets in each pile.

4. In the back of the store, Mr. Prescott packs 24 bell peppers equally into 8 bags. How many bell peppers are in each bag? Model the problem with both an array and a labeled tape diagram. Show each column as the number of bell peppers in each bag.

5. Olga saves $2 a week to buy a toy car. The car costs $16. How many weeks will it take her to save enough to buy the toy?

Lesson 11:   Model division as the unknown factor in multiplication using arrays and tape diagrams.

Name _____ Date _____

Ms. McCarty has 18 stickers. She puts 2 stickers on each homework paper and has no more left. How many homework papers does she have? Model the problem with both an array and a labeled tape diagram.

Lesson 11:   Model division as the unknown factor in multiplication using arrays and tape diagrams.

© 2018 Great Minds®. eureka-math.org

69

A chef arranges 4 rows of 3 red peppers on a tray. He adds 2 more rows of 3 yellow peppers.

How many peppers are there altogether?

_____

_____

_____

Read   **Draw**   **Write**

EUREKA
MATH

Lesson 12:   Interpret the quotient as the number of groups or the number of
              objects in each group using units of 2.

71

© 2018 Great Minds®. eureka-math.org

Name _____ Date _____

1. There are 8 birds at the pet store. Two birds are in each cage. Circle to show how many cages there are.

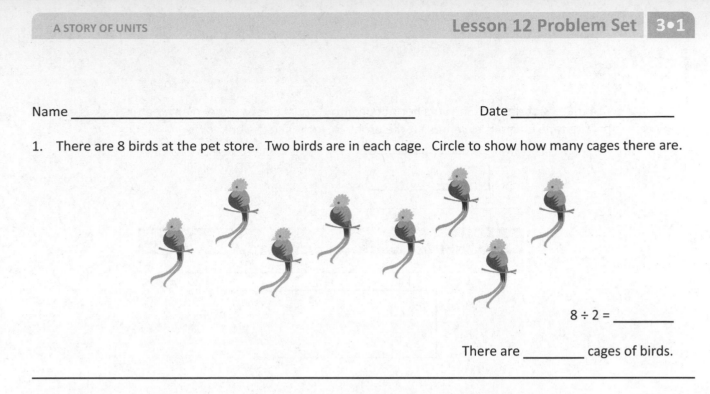

8 ÷ 2 = _____

There are _____ cages of birds.

2. The pet store sells 10 fish. They equally divide the fish into 5 bowls. Draw fish to find the number in each bowl.

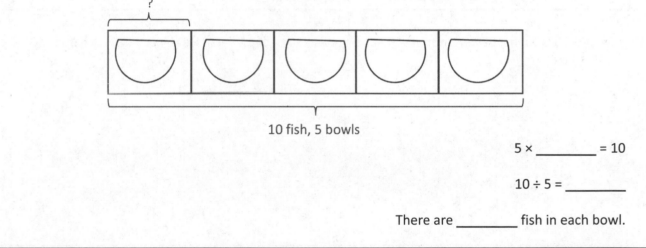

10 fish, 5 bowls

5 × _____ = 10

10 ÷ 5 = _____

There are _____ fish in each bowl.

3. Match.

Lesson 12: Interpret the quotient as the number of groups or the number of objects in each group using units of 2.

© 2018 Great Minds®. eureka-math.org

73

4. Laina buys 14 meters of ribbon. She cuts her ribbon into 2 equal pieces. How many meters long is each piece? Label the tape diagram to represent the problem, including the unknown.

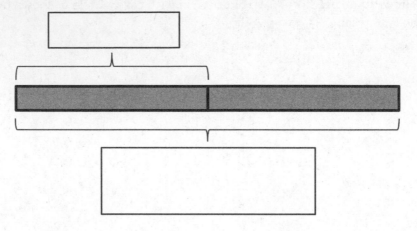

Each piece is _____ meters long.

_____

5. Roy eats 2 cereal bars every morning. Each box has a total of 12 bars. How many days will it take Roy to finish 1 box?

_____

6. Sarah and Esther equally share the cost of a present. The present costs $18. How much does Sarah pay?

Lesson 12:   Interpret the quotient as the number of groups or the number of objects in each group using units of 2.

© 2018 Great Minds®. eureka-math.org

Name _____   Date _____

There are 14 mints in 1 box. Cecilia eats 2 mints each day. How many days does it take Cecilia to eat 1 box of mints? Draw and label a tape diagram to solve.

It takes Cecilia _____ days to eat 1 box of mints.

Lesson 12:   Interpret the quotient as the number of groups or the number of
               objects in each group using units of 2.

© 2018 Great Minds®. eureka-math.org

75

Mark spends $16 on 2 video games. Each game costs the same amount. Find the cost of each game.

_____

_____

_____

_____

Read   Draw   Write

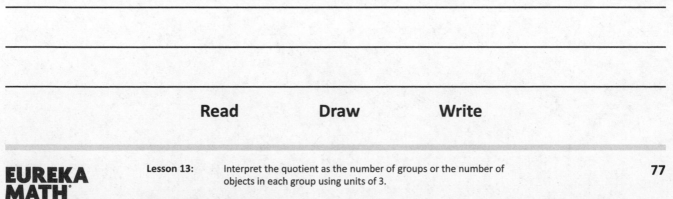

Lesson 13:   Interpret the quotient as the number of groups or the number of
objects in each group using units of 3.

© 2018 Great Minds®. eureka-math.org

77

Name _____   Date _____

1. Fill in the blanks to make true number sentences.

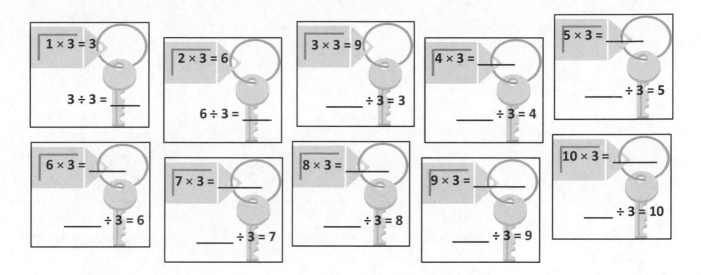

$1 \times 3 = 3$

$3 \div 3 = \underline{\quad}$

$2 \times 3 = 6$

$6 \div 3 = \underline{\quad}$

$3 \times 3 = 9$

$\underline{\quad} \div 3 = 3$

$4 \times 3 = \underline{\quad}$

$\underline{\quad} \div 3 = 4$

$5 \times 3 = \underline{\quad}$

$\underline{\quad} \div 3 = 5$

$6 \times 3 = \underline{\quad}$

$\underline{\quad} \div 3 = 6$

$7 \times 3 = \underline{\quad}$

$\underline{\quad} \div 3 = 7$

$8 \times 3 = \underline{\quad}$

$\underline{\quad} \div 3 = 8$

$9 \times 3 = \underline{\quad}$

$\underline{\quad} \div 3 = 9$

$10 \times 3 = \underline{\quad}$

$\underline{\quad} \div 3 = 10$

2. Mr. Lawton picks tomatoes from his garden. He divides the tomatoes into bags of 3.

    a. Circle to show how many bags he packs. Then, skip-count to show the total number of tomatoes.

    b. Draw and label a tape diagram to represent the problem.

_____ ÷ 3 = _____

Mr. Lawton packs _____ bags of tomatoes.

Lesson 13:   Interpret the quotient as the number of groups or the number of objects in each group using units of 3.

79

© 2018 Great Minds®. eureka-math.org

3. Camille buys a sheet of stamps that measures 15 centimeters long. Each stamp is 3 centimeters long. How many stamps does Camille buy? Draw and label a tape diagram to solve.

Camille buys _____ stamps.

4. Thirty third-graders go on a field trip. They are equally divided into 3 vans. How many students are in each van?

5. Some friends spend $24 altogether on frozen yogurt. Each person pays $3. How many people buy frozen yogurt?

Lesson 13:   Interpret the quotient as the number of groups or the number of objects in each group using units of 3.

Name _____   Date _____

1. Andrea uses 21 apple slices to decorate pies. She puts 3 slices on each pie. How many pies does Andrea make? Draw and label a tape diagram to solve.

2. There are 24 soccer players on the field. They form 3 equal teams. How many players are on each team?

Lesson 13:   Interpret the quotient as the number of groups or the number of
              objects in each group using units of 3.

81

© 2018 Great Minds®. eureka-math.org

Jackie buys 21 pizzas for a party. She places 3 pizzas on each table. How many tables are there?

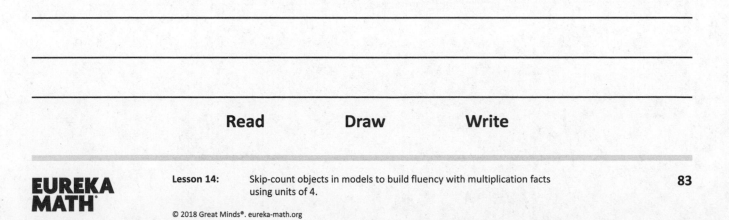

Read   **Draw**   **Write**

Lesson 14: Skip-count objects in models to build fluency with multiplication facts using units of 4.

83

Name _____ Date _____

1. Skip-count by fours. Match each answer to the appropriate expression.

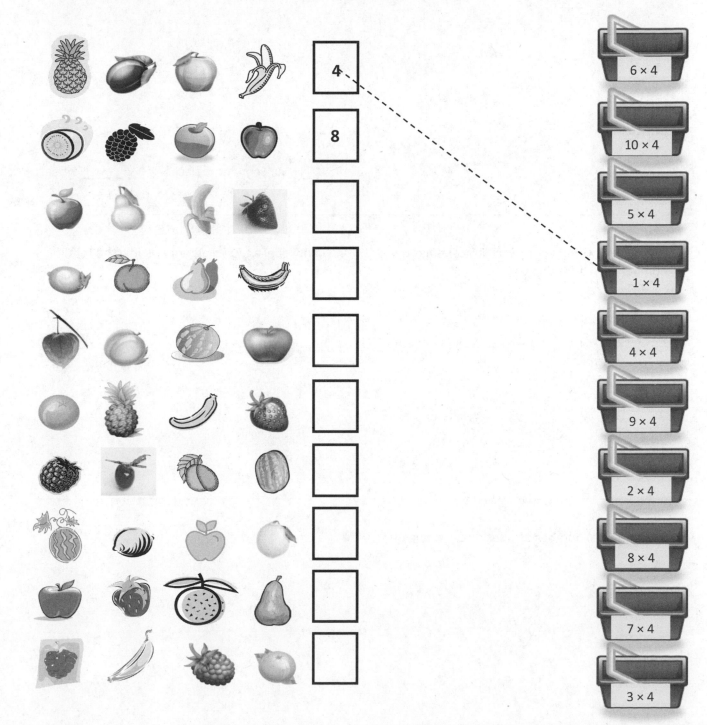

Lesson 14:   Skip-count objects in models to build fluency with multiplication facts using units of 4.

85

© 2018 Great Minds®. eureka-math.org

2. Mr. Schmidt replaces each of the 4 wheels on 7 cars. How many wheels does he replace? Draw and label a tape diagram to solve.

Mr. Schmidt replaces _____ wheels.

3. Trina makes 4 bracelets. Each bracelet has 6 beads. Draw and label a tape diagram to show the total number of beads Trina uses.

4. Find the total number of sides on 5 rectangles.

Lesson 14:   Skip-count objects in models to build fluency with multiplication facts using units of 4.

© 2018 Great Minds®. eureka-math.org

Name _____   Date _____

Arthur has 4 boxes of chocolates. Each box has 6 chocolates inside. How many chocolates does Arthur have altogether? Draw and label a tape diagram to solve.

Lesson 14:   Skip-count objects in models to build fluency with multiplication facts using units of 4.

© 2018 Great Minds®. eureka-math.org

87

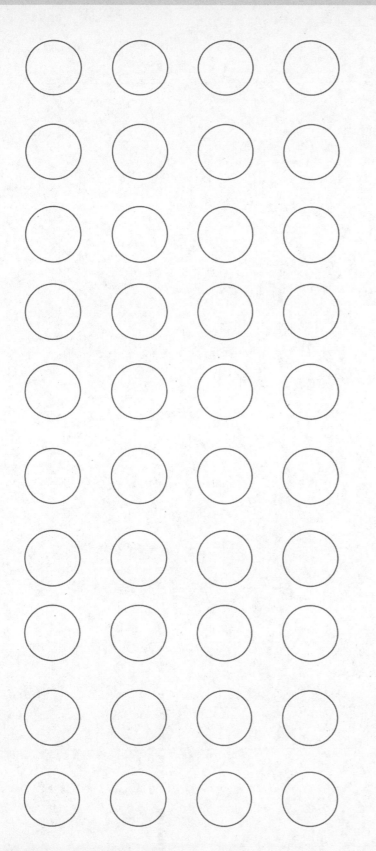

fours array

Lesson 14:   Skip-count objects in models to build fluency with multiplication facts using units of 4.

89

A cell phone is about 4 inches long. About how long are 9 cell phones laid end to end?

_____

_____

_____

_____

Read　　　　**Draw**　　　　**Write**

![EUREKA MATH] Lesson 15:　Relate arrays to tape diagrams to model the commutative property of multiplication.

91

© 2018 Great Minds®. eureka-math.org

Name _____ Date _____

1. Label the tape diagrams and complete the equations. Then, draw an array to represent the problems.

a.

4

$2 \times 4 =$ _____

2

$4 \times 2 =$ _____

b.

_____ $\times\ 4 =$ _____

$4 \times$ _____ $=$ _____

c.

_____ $\times$ _____ $= 28$

_____ $\times$ _____ $= 28$

Lesson 15:   Relate arrays to tape diagrams to model the commutative property of
             multiplication.

© 2018 Great Minds®. eureka-math.org

93

2. Draw and label 2 tape diagrams to model why the statement in the box is true.

$4 × 6 = 6 × 4$

3. Grace picks 4 flowers from her garden. Each flower has 8 petals. Draw and label a tape diagram to show how many petals there are in total.

4. Michael counts 8 chairs in his dining room. Each chair has 4 legs. How many chair legs are there altogether?

Lesson 15: Relate arrays to tape diagrams to model the commutative property of multiplication.

© 2018 Great Minds®. eureka-math.org

Name _____   Date _____

Draw and label 2 tape diagrams to show that 4 × 3 = 3 × 4. Use your diagrams to explain how you know the statement is true.

Lesson 15:   Relate arrays to tape diagrams to model the commutative property of multiplication.

© 2018 Great Minds®. eureka-math.org

95

Ms. Williams draws the array below to show the class seating chart. She sees the students in 4 rows of 7 when she teaches at Board 1. Use the commutative property to show how Ms. Williams sees the class when she teaches at Board 2.

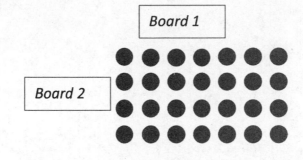

_____

_____

_____

_____

Read   Draw   Write

EUREKA MATH

Lesson 16: Use the distributive property as a strategy to find related multiplication facts.

© 2018 Great Minds®. eureka-math.org

97

Name _____   Date _____

1. Label the array. Then, fill in the blanks below to make true number sentences.

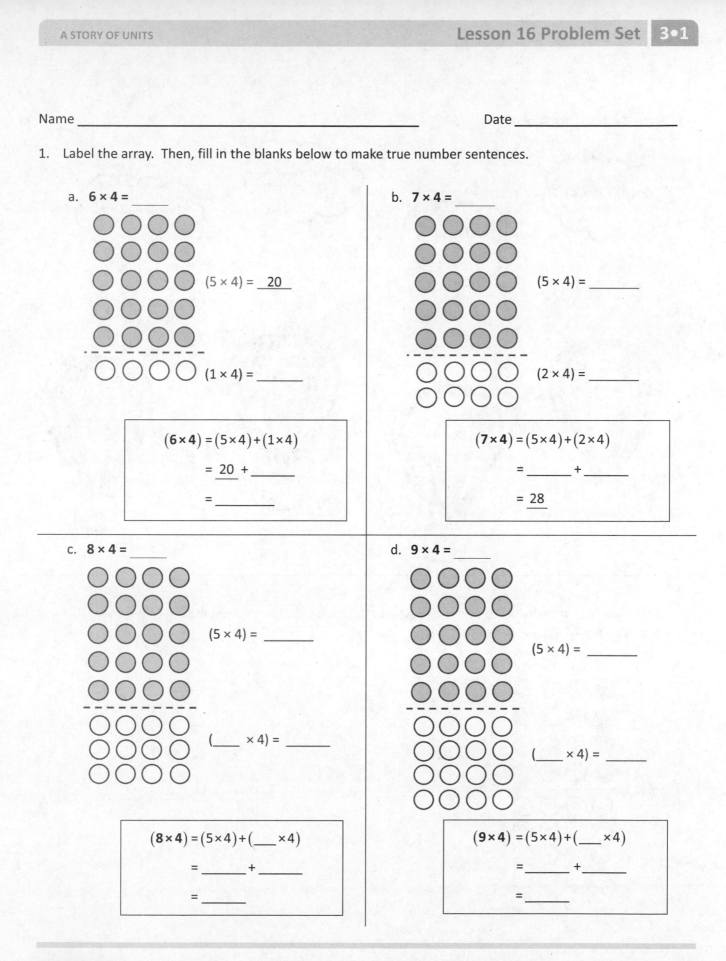

a. **6 × 4 =** _____

(5 × 4) = __20__

(1 × 4) = _____

$$(6 \times 4) = (5 \times 4) + (1 \times 4)$$

$$= \underline{20} + \underline{\hphantom{00}}$$

$$= \underline{\hphantom{0000}}$$

b. **7 × 4 =** _____

(5 × 4) = _____

(2 × 4) = _____

$$(7 \times 4) = (5 \times 4) + (2 \times 4)$$

$$= \underline{\hphantom{000}} + \underline{\hphantom{000}}$$

$$= \underline{28}$$

c. **8 × 4 =** _____

(5 × 4) = _____

(___ × 4) = _____

$$(8 \times 4) = (5 \times 4) + (\underline{\hphantom{0}} \times 4)$$

$$= \underline{\hphantom{000}} + \underline{\hphantom{000}}$$

$$= \underline{\hphantom{000}}$$

d. **9 × 4 =** _____

(5 × 4) = _____

(___ × 4) = _____

$$(9 \times 4) = (5 \times 4) + (\underline{\hphantom{0}} \times 4)$$

$$= \underline{\hphantom{000}} + \underline{\hphantom{000}}$$

$$= \underline{\hphantom{000}}$$

2. Match the equal expressions.

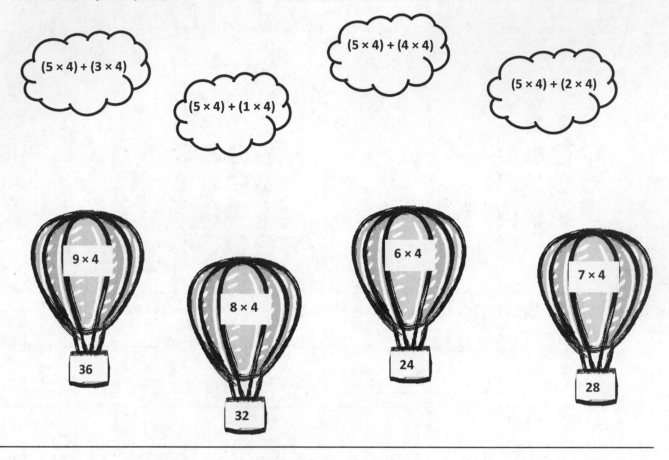

$(5 \times 4) + (3 \times 4)$

$(5 \times 4) + (1 \times 4)$

$(5 \times 4) + (4 \times 4)$

$(5 \times 4) + (2 \times 4)$

9×4

8×4

6×4

7×4

36

32

24

28

3. Nolan draws the array below to find the answer to the multiplication expression 10×4. He says, "10×4 is just double 5×4." Explain Nolan's strategy.

Lesson 16:   Use the distributive property as a strategy to find related multiplication facts.

EUREKA MATH

Name _____ Date _____

Destiny says, "I can use 5 × 4 to find the answer to 7 × 4." Use the array below to explain Destiny's strategy using words and numbers.

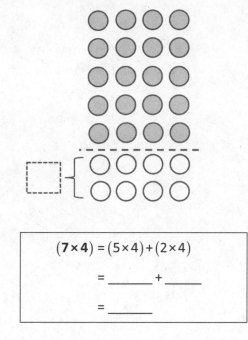

$$(\mathbf{7 \times 4}) = (5 \times 4) + (2 \times 4)$$

$$= \rule{1.5cm}{0.4pt} + \rule{1.5cm}{0.4pt}$$

$$= \rule{1.5cm}{0.4pt}$$

Lesson 16: Use the distributive property as a strategy to find related multiplication facts.

© 2018 Great Minds®. eureka-math.org

101

Mrs. Peacock bought 4 packs of yogurt. She had exactly enough to give each of her 24 students 1 yogurt cup. How many yogurt cups are there in 1 pack?

_____

_____

_____

_____

Read   **Draw**   **Write**

Name _____   Date _____

1. Use the array to complete the related equations.

$1 \times 4 = \underline{\quad 4 \quad}$ $\underline{\quad 4 \quad} \div 4 = 1$

$2 \times 4 = \underline{\qquad}$ $\underline{\qquad} \div 4 = 2$

$\underline{\qquad} \times 4 = 12$ $12 \div 4 = \underline{\qquad}$

$\underline{\qquad} \times 4 = 16$ $16 \div 4 = \underline{\qquad}$

$\underline{\qquad} \times \underline{\qquad} = 20$ $20 \div \underline{\qquad} = \underline{\qquad}$

$\underline{\qquad} \times \underline{\qquad} = 24$ $24 \div \underline{\qquad} = \underline{\qquad}$

$\underline{\qquad} \times 4 = \underline{\qquad}$ $\underline{\qquad} \div 4 = \underline{\qquad}$

$\underline{\qquad} \times 4 = \underline{\qquad}$ $\underline{\qquad} \div 4 = \underline{\qquad}$

$\underline{\qquad} \times \underline{\qquad} = \underline{\qquad}$ $\underline{\qquad} \div \underline{\qquad} = \underline{\qquad}$

$\underline{\qquad} \times \underline{\qquad} = \underline{\qquad}$ $\underline{\qquad} \div \underline{\qquad} = \underline{\qquad}$

EUREKA MATH®

2. The baker packs 36 bran muffins in boxes of 4. Draw and label a tape diagram to find the number of boxes he packs.

3. The waitress arranges 32 glasses into 4 equal rows. How many glasses are in each row?

4. Janet paid $28 for 4 notebooks. Each notebook costs the same amount. What is the cost of 2 notebooks?

Lesson 17: Model the relationship between multiplication and division.

Name _____   Date _____

1. Mr. Thomas organizes 16 binders into stacks of 4. How many stacks does he make? Draw and label a number bond to solve.

2. The chef uses 28 avocados to make 4 batches of guacamole. How many avocados are in 2 batches of guacamole? Draw and label a tape diagram to solve.

A parking structure has 10 levels. There are 3 cars parked on each level. How many cars are parked in the structure?

_____

_____

_____

_____

Read   Draw   Write

Name _____ Date _____

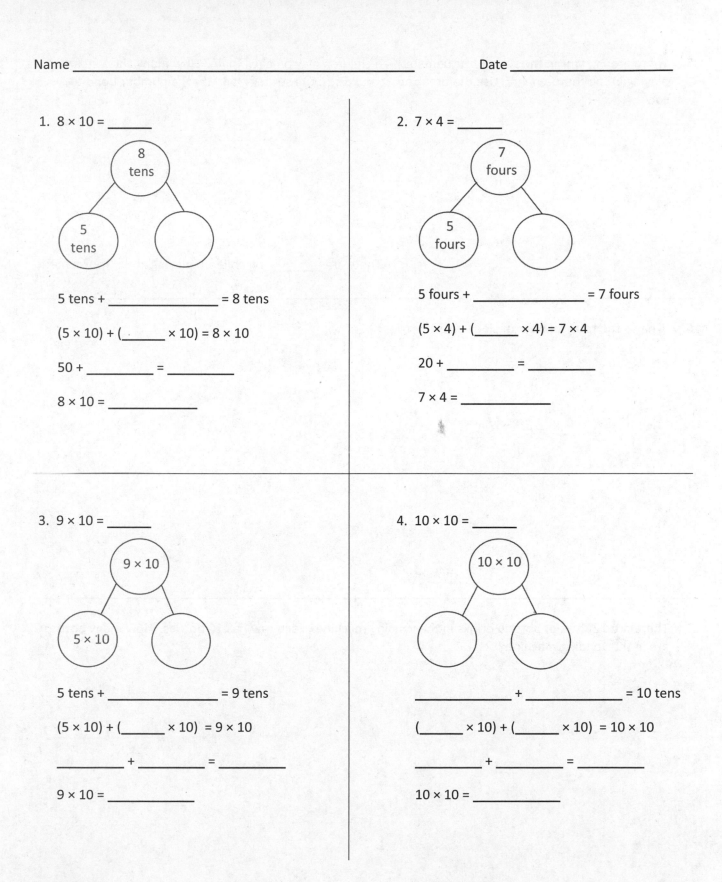

1. 8 × 10 = _____

    8 tens

    5 tens

    5 tens + _____ = 8 tens

    (5 × 10) + (_____ × 10) = 8 × 10

    50 + _____ = _____

    8 × 10 = _____

2. 7 × 4 = _____

    7 fours

    5 fours

    5 fours + _____ = 7 fours

    (5 × 4) + (_____ × 4) = 7 × 4

    20 + _____ = _____

    7 × 4 = _____

3. 9 × 10 = _____

    9 × 10

    5 × 10

    5 tens + _____ = 9 tens

    (5 × 10) + (_____ × 10) = 9 × 10

    _____ + _____ = _____

    9 × 10 = _____

4. 10 × 10 = _____

    10 × 10

    _____ + _____ = 10 tens

    (_____ × 10) + (_____ × 10) = 10 × 10

    _____ + _____ = _____

    10 × 10 = _____

EUREKA MATH

5. There are 7 teams in the soccer tournament. Ten children play on each team. How many children are playing in the tournament? Use the break apart and distribute strategy, and draw a number bond to solve.

There are _____ children playing in the tournament.

6. What is the total number of sides on 8 triangles?

7. There are 12 rows of bottled drinks in the vending machine. Each row has 10 bottles. How many bottles are in the vending machine?

Name _____   Date _____

Dylan used the break apart and distribute strategy to solve a multiplication problem. Look at his work below, write the multiplication problem Dylan solved, and complete the number bond.

Dylan's work:

$(5 \times 4) + (1 \times 4) =$

$20 + 4 = 24$

```
        ( )
       /   \
   (5 × 4) ( )
```

_____ × _____ = _____

EUREKA MATH

Henrietta works in a shoe store. She uses 2 shoelaces to lace each pair of shoes. She has a total of 24 laces. How many pairs of shoes can Henrietta lace?

_____

_____

_____

_____

Read   **Draw**   **Write**

Name _____ Date _____

1. Label the array. Then, fill in the blanks to make true number sentences.

a. 36 ÷ 3 = _____

(30 ÷ 3) = _____

(6 ÷ 3) = _____

$$(36 \div 3) = (30 \div 3) + (6 \div 3)$$

$$= \underline{10} + \underline{}$$

$$= \underline{12}$$

b. 25 ÷ 5 = _____

(20 ÷ 5) = _4_

(5 ÷ 5) = _____

$$(25 \div 5) = (20 \div 5) + (5 \div 5)$$

$$= \underline{4} + \underline{}$$

$$= \underline{}$$

c. 28 ÷ 4 = _____

(20 ÷ 4) = _____

(___ ÷ 4) = ___

$$(28 \div 4) = (20 \div 4) + (\underline{} \div 4)$$

$$= \underline{} + \underline{}$$

$$= \underline{}$$

d. 32 ÷ 4 = _____

(___ ÷ 4) = _____

(___ ÷ 4) = ___

$$(32 \div 4) = (\underline{} \div 4) + (\underline{} \div 4)$$

$$= \underline{} + \underline{}$$

$$= \underline{}$$

2. Match the equal expressions.

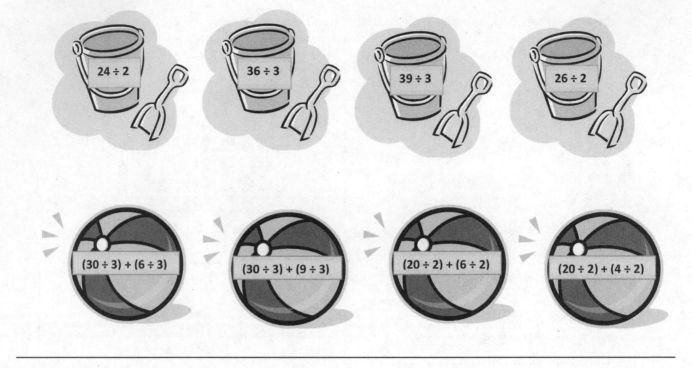

3. Nell draws the array below to find the answer to 24 ÷ 2. Explain Nell's strategy.

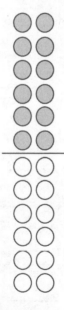

Lesson 19:   Apply the distributive property to decompose units.

Name _____   Date _____

Complete the equations below to solve 22 ÷ 2 = _____.

$(20 \div 2) =$ _____

$(\underline{\hspace{1cm}} \div 2) =$ _____

$(22 \div 2) = (20 \div 2) + (\underline{\hspace{1cm}} \div 2)$

$= \underline{\hspace{1cm}} + \underline{\hspace{1cm}}$

$= \underline{\hspace{1cm}}$

Red, orange, and blue scarves are on sale for $4 each. Nina buys 2 scarves of each color. How much does she spend altogether?

_____

_____

_____

_____

Read   **Draw**   **Write**

EUREKA
MATH

Lesson 20:   Solve two-step word problems involving multiplication and division, and assess the reasonableness of answers.

121

© 2018 Great Minds®. eureka-math.org

Name _____   Date _____

1. Ted buys 3 books and a magazine at the book store. Each book costs $8. A magazine costs $4.

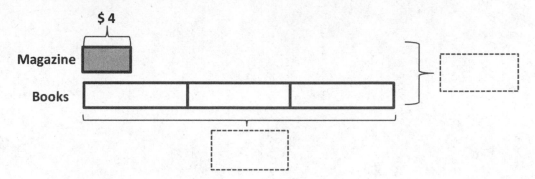

    a. What is the total cost of the books?

    b. How much does Ted spend altogether?

2. Seven children share 28 silly bands equally.

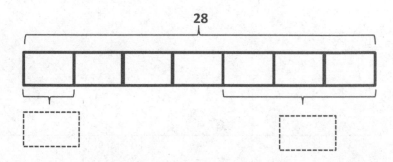

    a. How many silly bands does each child get?

    b. How many silly bands do 3 children get?

 EUREKA MATH   Lesson 20:   Solve two-step word problems involving multiplication and division,   123
                                and assess the reasonableness of answers.

© 2018 Great Minds®. eureka-math.org

3. Eighteen cups are equally packed into 6 boxes. Two boxes of cups break. How many cups are unbroken?

---

4. There are 25 blue balloons and 15 red balloons at a party. Five children are given an equal number of each color balloon. How many blue and red balloons does each child get?

---

5. Twenty-seven pears are packed in bags of 3. Five bags of pears are sold. How many bags of pears are left?

Lesson 20: Solve two-step word problems involving multiplication and division, and assess the reasonableness of answers.

Name _____   Date _____

1. Thirty-two jelly beans are shared by 8 students.

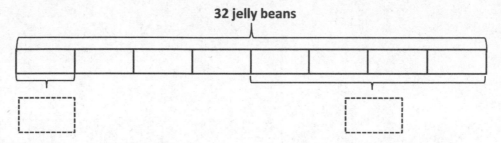

    a. How many jelly beans will each student get?

    b. How many jelly beans will 4 students get?

2. The teacher has 30 apple slices and 20 pear slices. Five children equally share all of the fruit slices. How many fruit slices does each child get?

Lesson 20:   Solve two-step word problems involving multiplication and division, and assess the reasonableness of answers.

125

© 2018 Great Minds®. eureka-math.org

There are 4 boxes with 6 binders in each one. Three brothers share the binders. How many binders does each brother get?

Read   **Draw**   **Write**

EUREKA MATH

Lesson 21:   Solve two-step word problems involving all four operations, and assess
               the reasonableness of answers.

127

© 2018 Great Minds®. eureka-math.org

Name _____   Date _____

1. Jason earns $6 per week for doing all his chores. On the fifth week, he forgets to take out the trash, so he only earns $4. Write and solve an equation to show how much Jason earns in 5 weeks.

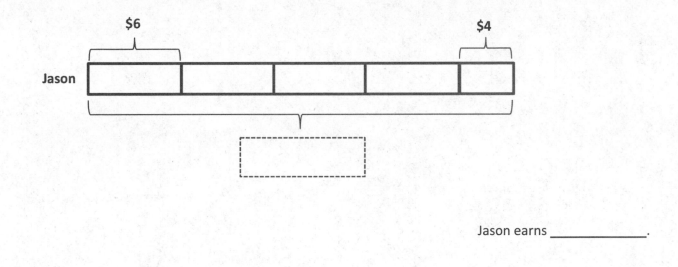

Jason earns _____.

_____

2. Miss Lianto orders 4 packs of 7 markers. After passing out 1 marker to each student in her class, she has 6 left. Label the tape diagram to find how many students are in Miss Lianto's class.

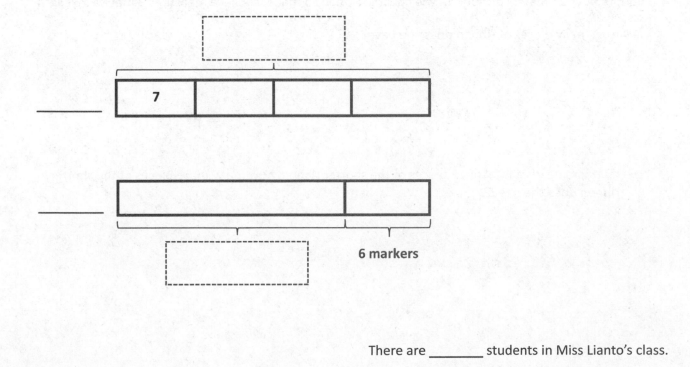

There are _____ students in Miss Lianto's class.

Lesson 21: Solve two-step word problems involving all four operations, and assess the reasonableness of answers.

129

© 2018 Great Minds®. eureka-math.org

3. Orlando buys a box of 18 fruit snacks. Each box comes with an equal number of strawberry-, cherry-, and grape-flavored snacks. He eats all of the grape-flavored snacks. Draw and label a tape diagram to find how many fruit snacks he has left.

---

4. Eudora buys 21 meters of ribbon. She cuts the ribbon so that each piece measures 3 meters in length.

   a. How many pieces of ribbon does she have?

   b. If Eudora needs a total of 12 pieces of the shorter ribbon, how many more pieces of the shorter ribbon does she need?

      Lesson 21:   Solve two-step word problems involving all four operations, and assess the reasonableness of answers.

Name _____   Date _____

Ms. Egeregor buys 27 books for her classroom library. She buys an equal number of fiction, nonfiction, and poetry books. She shelves all of the poetry books first. Draw and label a tape diagram to show how many books Ms. Egeregor has left to shelve.

EUREKA
MATH®

Lesson 21:   Solve two-step word problems involving all four operations, and assess
                the reasonableness of answers.

© 2018 Great Minds®. eureka-math.org

131

Ms. Bower helps her kindergartners tie their shoes. It takes her 5 seconds to tie 1 shoe. How many seconds does it take Ms. Bower to tie 8 shoes?

_____

_____

_____

_____

Read   **Draw**   **Write**

Lesson 1:   Explore time as a continuous measurement using a stopwatch.

EUREKA MATH

135

© 2018 Great Minds®. eureka-math.org

Name _____   Date _____

1. Use a stopwatch. How long does it take you to snap your fingers 10 times?

    It takes _____ to snap 10 times.

2. Use a stopwatch. How long does it take to write every whole number from 0 to 25?

    It takes _____ to write every whole number from 0 to 25.

3. Use a stopwatch. How long does it take you to name 10 animals? Record them below.

    It takes _____ to name 10 animals.

4. Use a stopwatch. How long does it take you to write 7 × 8 = 56 fifteen times? Record the time below.

    It takes _____ to write 7 × 8 = 56 fifteen times.

EUREKA MATH

Lesson 1:   Explore time as a continuous measurement using a stopwatch.

137

© 2018 Great Minds®. eureka-math.org

5. Work with your group. Use a stopwatch to measure the time for each of the following activities.

Activity		Time
Write your full name.		_____ seconds
Do 20 jumping jacks.		
Whisper count by twos from 0 to 30.		
Draw 8 squares.		
Skip-count out loud by fours from 24 to 0.		
Say the names of your teachers from Kindergarten to Grade 3.		

6. 100 meter relay: Use a stopwatch to measure and record your team's times.

Name	Time
	Total time:

     Lesson 1:   Explore time as a continuous measurement using a stopwatch.

Name _____   Date _____

The table to the right shows how much time it takes each of the 5 students to do 15 jumping jacks.

Maya	16 seconds
Riley	15 seconds
Jake	14 seconds
Nicholas	15 seconds
Adeline	17 seconds

    a.   Who finished 15 jumping jacks the fastest?

    b.   Who finished their jumping jacks in the exact same amount of time?

    c.   How many seconds faster did Jake finish than Adeline?

Christine has 12 math problems for homework. It takes her 5 minutes to complete each problem.

How many minutes does it take Christine to finish all 12 problems?

Read **Draw** **Write**

Lesson 2: Relate skip-counting by fives on the clock and telling time to a continuous measurement model, the number line.

141

Name _____   Date _____

1. Follow the directions to label the number line below.

   a. Ingrid gets ready for school between 7:00 a.m. and 8:00 a.m. Label the first and last tick marks as 7:00 a.m. and 8:00 a.m.

   b. Each interval represents 5 minutes. Count by fives starting at 0, or 7:00 a.m. Label each 5-minute interval below the number line up to 8:00 a.m.

   c. Ingrid starts getting dressed at 7:10 a.m. Plot a point on the number line to represent this time. Above the point, write D.

   d. Ingrid starts eating breakfast at 7:35 a.m. Plot a point on the number line to represent this time. Above the point, write E.

   e. Ingrid starts brushing her teeth at 7:40 a.m. Plot a point on the number line to represent this time. Above the point, write T.

   f. Ingrid starts packing her lunch at 7:45 a.m. Plot a point on the number line to represent this time. Above the point, write L.

   g. Ingrid starts waiting for the bus at 7:55 a.m. Plot a point on the number line to represent this time. Above the point, write W.

EUREKA MATH   Lesson 2:   Relate skip-counting by fives on the clock and telling time to a   143
                                    continuous measurement model, the number line.

© 2018 Great Minds®. eureka-math.org

2. Label every 5 minutes below the number line shown. Draw a line from each clock to the point on the number line which shows its time. Not all of the clocks have matching points.

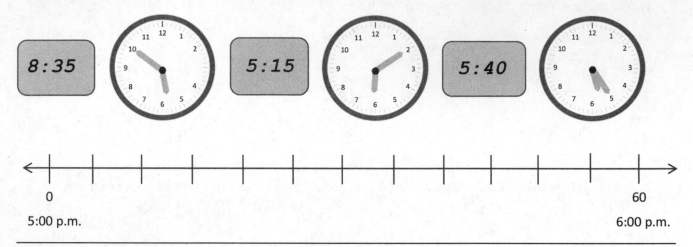

0

5:00 p.m.

60

6:00 p.m.

3. Noah uses a number line to locate 5:45 p.m. Each interval is 5 minutes. The number line shows the hour from 5 p.m. to 6 p.m. Label the number line below to show his work.

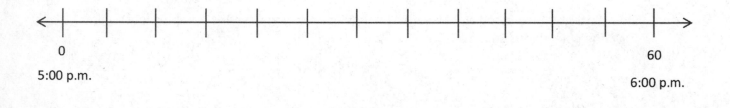

0

5:00 p.m.

60

6:00 p.m.

4. Tanner tells his little brother that 11:25 p.m. comes after 11:20 a.m. Do you agree with Tanner? Why or why not?

Lesson 2:   Relate skip-counting by fives on the clock and telling time to a continuous measurement model, the number line.

© 2018 Great Minds®. eureka-math.org

Name _____   Date _____

The number line below shows a math class that begins at 10:00 a.m. and ends at 11:00 a.m. Use the number line to answer the following questions.

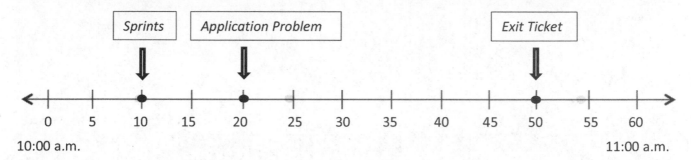

10:00 a.m.   11:00 a.m.

    a.   What time do Sprints begin?

    b.   What time do students begin the Application Problem?

    c.   What time do students work on the Exit Ticket?

    d.   How long is math class?

Lesson 2:   Relate skip-counting by fives on the clock and telling time to a
              continuous measurement model, the number line.

145

© 2018 Great Minds®. eureka-math.org

tape diagram

Lesson 2:   Relate skip-counting by fives on the clock and telling time to a
              continuous measurement model, the number line.

147

© 2018 Great Minds®. eureka-math.org

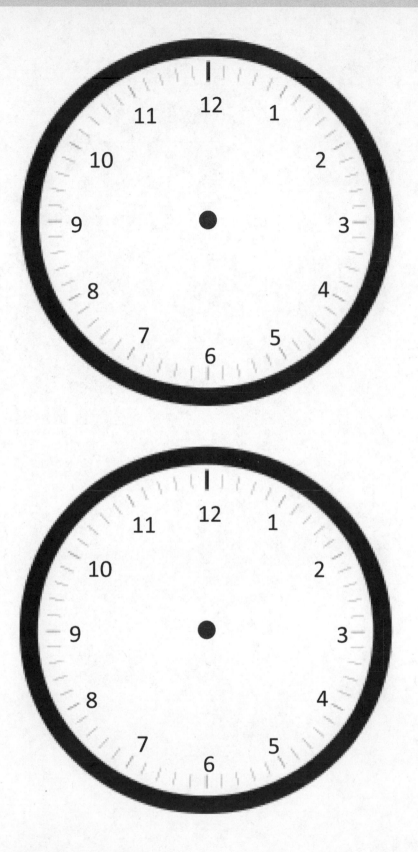

two clocks

Lesson 2:   Relate skip-counting by fives on the clock and telling time to a
continuous measurement model, the number line.

There are 12 tables in the cafeteria. Five students sit at each of the first 11 tables. Three students sit at the last table. How many students are sitting at the 12 tables in the cafeteria?

_____

_____

_____

_____

Read   **Draw**   **Write**

EUREKA MATH

Lesson 3:   Count by fives and ones on the number line as a strategy to tell time to
             the nearest minute on the clock.

© 2018 Great Minds®. eureka-math.org

151

Name _____   Date _____

1. Plot a point on the number line for the times shown on the clocks below. Then, draw a line to match the clocks to the points.

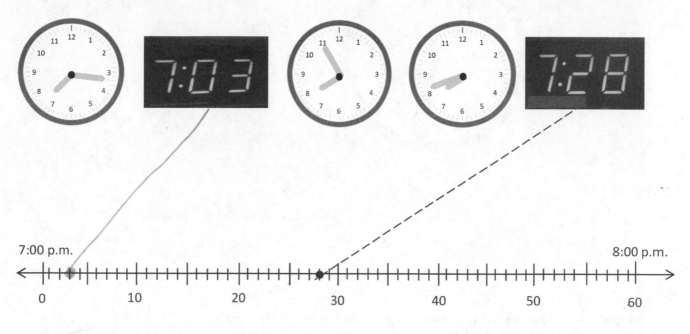

7:00 p.m.   8:00 p.m.

0   10   20   30   40   50   60

2. Jessie woke up this morning at 6:48 a.m. Draw hands on the clock below to show what time Jessie woke up.

3. Mrs. Barnes starts teaching math at 8:23 a.m. Draw hands on the clock below to show what time Mrs. Barnes starts teaching math.

Lesson 3:   Count by fives and ones on the number line as a strategy to tell time to the nearest minute on the clock.

153

© 2018 Great Minds®. eureka-math.org

4. The clock shows what time Rebecca finishes her homework. What time does Rebecca finish her homework?

Rebecca finishes her homework at _5:27_.

5. The clock below shows what time Mason's mom drops him off for practice.

   a. What time does Mason's mom drop him off?

   3:56

   b. Mason's coach arrived 11 minutes before Mason. What time did Mason's coach arrive?

   3:56
   − 11
   ——————
   3:45

Lesson 3:   Count by fives and ones on the number line as a strategy to tell time to the nearest minute on the clock.

© 2018 Great Minds®. eureka-math.org

EUREKA MATH

Name _____   Date _____

The clock shows what time Jason gets to school in the morning.

Arrival at School 8:03

a. What time does Jason get to school?

    Jason arrives a 8:03

b. The first bell rings at 8:23 a.m. Draw hands on the clock
    to show when the first bell rings.

    Alrt b

First Bell Rings

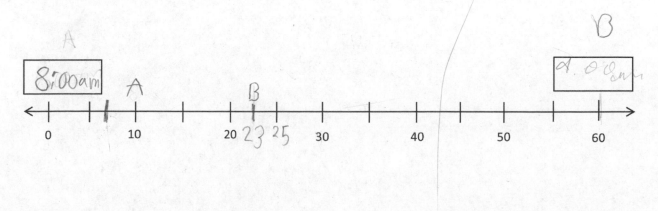

c. Label the first and last tick marks 8:00 a.m. and 9:00 a.m. Plot a point to show when Jason arrives at
    school. Label it *A*. Plot a point on the line when the first bell rings and label it *B*.

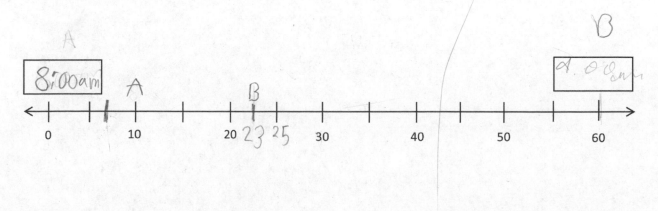

Lesson 3:   Count by fives and ones on the number line as a strategy to tell time to
             the nearest minute on the clock.

155

© 2018 Great Minds®. eureka-math.org

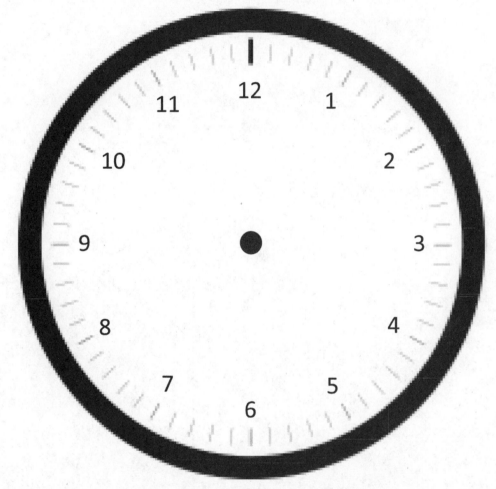

clock

Lesson 3:   Count by fives and ones on the number line as a strategy to tell time to
           the nearest minute on the clock.

157

© 2018 Great Minds®. eureka-math.org

Patrick and Lilly start their chores at 5:00 p.m. The clock shows what time Lilly finishes. The number line shows what time Patrick finishes. Who finishes first? Explain how you know. Solve the problem without drawing a number line.

Lilly

Patrick

5:00 p.m.   6:00 p.m.

0   60

_____

_____

_____

_____

Read   Draw   Write

EUREKA MATH

Name _____   Date _____

Use a number line to answer Problems 1 through 5.

1. Cole starts reading at 6:23 p.m. He stops at 6:49 p.m. How many minutes does Cole read?   6:23 →3 6:49

    6:23   6:49
    23, 24, 25, 26, 27 6:23 29 30
                    026   Cole reads for __26__ minutes.

2. Natalie finishes piano practice at 2:45 p.m. after practicing for 37 minutes. What time did Natalie's practice start?

    3 15
    4 5
    - 3 7
    0 8   Natalie's practice started at __2:08__ p.m.

3. Genevieve works on her scrapbook from 11:27 a.m. to 11:58 a.m. How many minutes does she work on her scrapbook?

    (27) 11:27 20+10+10+10=50   7+1=8   58
                                                       31 - 27
                                                            31
                           Genevieve works on her scrapbook for __31__ minutes.

4. Nate finishes his homework at 4:47 p.m. after working on it for 38 minutes. What time did Nate start his homework?

    3 17
    4 7
    - 3 8
    0 9   Nate started his homework at __4:09__ p.m.

5. Andrea goes fishing at 9:03 a.m. She fishes for 49 minutes. What time is Andrea done fishing?

    1
    4 9
    + 3
    5 2   Andrea is done fishing at __9:52__ a.m.

6. Dion walks to school. The clocks below show when he leaves his house and when he arrives at school. How many minutes does it take Dion to walk to school?

Dion leaves his house: 7:37   Dion arrives at school: 7:56

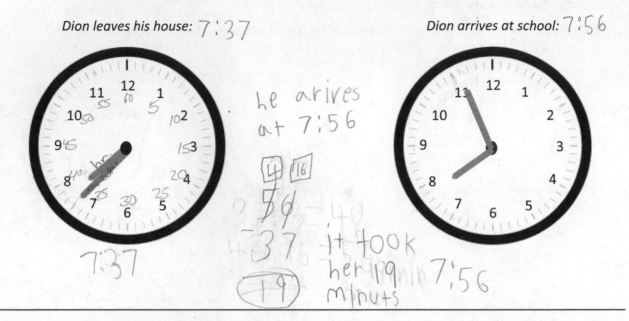

he arives
at 7:56

[4] [16]

5̶6̶ = 40

37

it took
her 19 min
minuts

(19)

7:37   7:56

7. Sydney cleans her room for 45 minutes. She starts at 11:13 a.m. What time does Sydney finish cleaning her room?

13
+45

58

She finish cleaning
at 11:58

8. The third-grade chorus performs a musical for the school. The musical lasts 42 minutes. It ends at 1:59 p.m. What time did the musical start?

59
-42

17

it Starts
at 1:17

Lesson 4:   Solve word problems involving time intervals within 1 hour by counting backward and forward using the number line and clock.

© 2018 Great Minds®. eureka-math.org

Name _____   Date _____

Independent reading time starts at 1:34 p.m. It ends at 1:56 p.m.

1. Draw the start time on the clock below.

2. Draw the end time on the clock below.

3. How many minutes does independent reading time last?

Lesson 4:   Solve word problems involving time intervals within 1 hour by counting
              backward and forward using the number line and clock.

163

© 2018 Great Minds®. eureka-math.org

number line

Lesson 4: Solve word problems involving time intervals within 1 hour by counting backward and forward using the number line and clock.

165

Carlos gets to class at 9:08 a.m. He has to write down homework assignments and complete morning work before math begins at 9:30 a.m. How many minutes does Carlos have to complete his tasks before math begins?

_____

_____

_____

_____

Read   **Draw**   **Write**

EUREKA MATH

Lesson 5:   Solve word problems involving time intervals within 1 hour by adding and subtracting on the number line.

167

© 2018 Great Minds®. eureka-math.org

Name _____ Date _____

1. Cole read his book for 25 minutes yesterday and for 28 minutes today. How many minutes did Cole read altogether? Model the problem on the number line, and write an equation to solve.

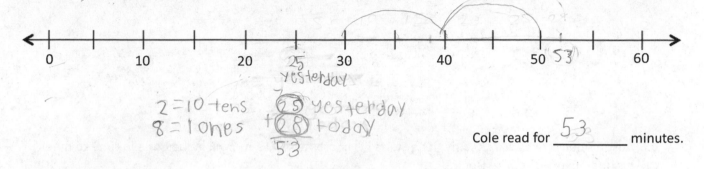

2 = 10 tens   25 yesterday
8 = 1 ones   + 28 today
                    53

Cole read for ____53____ minutes.

2. Tessa spends 34 minutes washing her dog. It takes her 12 minutes to shampoo and rinse and the rest of the time to get the dog in the bathtub! How many minutes does Tessa spend getting her dog in the bathtub? Draw a number line to model the problem, and write an equation to solve.

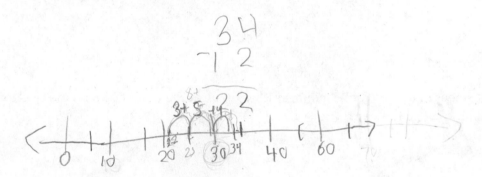

3. Tessa walks her dog for 47 minutes. Jeremiah walks his dog for 30 minutes. How many more minutes does Tessa walk her dog than Jeremiah?

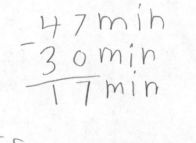

47 min
- 30 min
  17 min

Tessa walks her dog 17 more minutes then Jeremiha.

Lesson 5: Solve word problems involving time intervals within 1 hour by adding and subtracting on the number line.

169

© 2018 Great Minds®. eureka-math.org

4. a. It takes Austin 4 minutes to take out the garbage, 12 minutes to wash the dishes, and 13 minutes to mop the kitchen floor. How long does it take Austin to do his chores?

    b. Austin's bus arrives at 7:55 a.m. If he starts his chores at 7:30 a.m., will he be done in time to meet his bus? Explain your reasoning.

5. Gilberto's cat sleeps in the sun for 23 minutes. It wakes up at the time shown on the clock below. What time did the cat go to sleep?

Lesson 5:   Solve word problems involving time intervals within 1 hour by adding and subtracting on the number line.

© 2018 Great Minds®. eureka-math.org

Name _____   Date _____

Michael spends 19 minutes on his math homework and 17 minutes on his science homework.

How many minutes does Michael spend doing his homework?

Model the problem on the number line, and write an equation to solve.

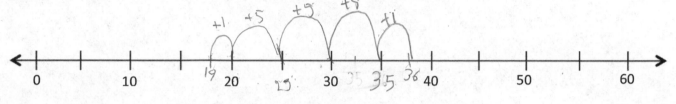

19 min
+17 min
36

Michael spends ___36___ minutes on his homework.

Lesson 5:   Solve word problems involving time intervals within 1 hour by adding
and subtracting on the number line.

171

© 2018 Great Minds®. eureka-math.org

Name _____ Date _____

1. Illustrate and describe the process of making a 1-kilogram weight.

2. Illustrate and describe the process of decomposing 1 kilogram into groups of 100 grams.

3. Illustrate and describe the process of decomposing 100 grams into groups of 10 grams.

Lesson 6:   Build and decompose a kilogram to reason about the size and weight of 1 kilogram, 100 grams, 10 grams, and 1 gram.

173

© 2018 Great Minds®. eureka-math.org

4. Illustrate and describe the process of decomposing 10 grams into groups of 1 gram.

5. Compare the two place value charts below. How does today's exploration using kilograms and grams relate to your understanding of place value?

1 kilogram	100 grams	10 grams	1 gram

Thousands	Hundreds	Tens	Ones

Lesson 6:   Build and decompose a kilogram to reason about the size and weight of 1 kilogram, 100 grams, 10 grams, and 1 gram.

Name _____   Date _____

Ten bags of sugar weigh 1 kilogram. How many grams does each bag of sugar weigh?

Lesson 6:   Build and decompose a kilogram to reason about the size and weight
            of 1 kilogram, 100 grams, 10 grams, and 1 gram.

© 2018 Great Minds®. eureka-math.org

175

Justin put a 1-kilogram bag of flour on one side of a pan balance. How many 100-gram bags of flour does he need to put on the other pan to balance the scale?

_____

_____

_____

_____

Read   **Draw**   **Write**

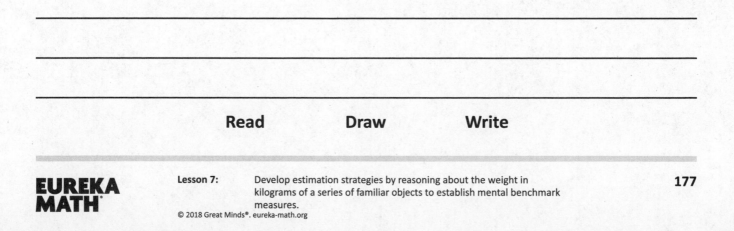

Name _____   Date _____

Work with a partner. Use the corresponding weights to estimate the weight of objects in the classroom.
Then, check your estimate by weighing on a scale.

A.

Objects that Weigh About **1 Kilogram**	Actual Weight

B.

Objects that Weigh About **100 Grams**	Actual Weight

C.

Objects that Weigh About **10 Grams**	Actual Weight

D.

Objects that Weigh About **1 Gram**	Actual Weight

Lesson 7:   Develop estimation strategies by reasoning about the weight in
kilograms of a series of familiar objects to establish mental benchmark
measures.

© 2018 Great Minds®. eureka-math.org

179

E. Circle the correct unit of weight for each estimation.

1. A box of cereal weighs about 350 (grams / kilograms).

2. A watermelon weighs about 3 (grams / kilograms).

3. A postcard weighs about 6 (grams / kilograms).

4. A cat weighs about 4 (grams / kilograms).

5. A bicycle weighs about 15 (grams / kilograms).

6. A lemon weighs about 58 (grams / kilograms).

F. During the exploration, Derrick finds that his bottle of water weighs the same as a 1-kilogram bag of rice. He then exclaims, "Our class laptop weighs the same as 2 bottles of water!" How much does the laptop weigh in kilograms? Explain your reasoning.

G. Nessa tells her brother that 1 kilogram of rice weighs the same as 10 bags containing 100 grams of beans each. Do you agree with her? Explain why or why not.

Lesson 7:   Develop estimation strategies by reasoning about the weight in kilograms of a series of familiar objects to establish mental benchmark measures.
© 2018 Great Minds®. eureka-math.org

Name _____   Date _____

1. Read and write the weights below. Write the word *kilogram* or *gram* with the measurement.

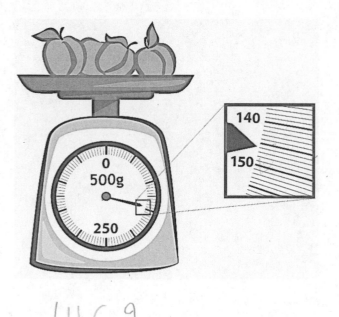

146g

12 kg

2. Circle the correct unit of weight for each estimation.

   a. An orange weighs about 200 (grams / kilograms).

   b. A basketball weighs about 624 (grams / kilograms).

   c. A brick weighs about 2 (grams / kilograms).

   d. A small packet of sugar weighs about 4 (grams / kilograms).

   e. A tiger weighs about 190 (grams / kilograms).

Name _____   Date _____

1. Tim goes to the market to buy fruits and vegetables. He weighs some string beans and some grapes.

List the weights for both the string beans and grapes.

The string beans weigh ___464___ grams.

The grapes weigh ___355___ grams.

2. Use tape diagrams to model the following problems. Keiko and her brother Jiro get weighed at the doctor's office. Keiko weighs 35 kilograms, and Jiro weighs 43 kilograms.

a. What is Keiko and Jiro's total weight?

$$\begin{array}{r} 43 \\ +35 \\ \hline 78 \end{array}$$

Keiko and Jiro weigh ___78___ kilograms.

b. How much heavier is Jiro than Keiko?

$$\begin{array}{r} 4\overset{3}{3} \\ -35 \\ \hline 08 \end{array}$$

Jiro is ___8___ kilograms heavier than Keiko.

Lesson 8:   Solve one-step word problems involving metric weights within 100 and estimate to reason about solutions.

183

© 2018 Great Minds®. eureka-math.org

3. Jared estimates that his houseplant is as heavy as a 5-kilogram bowling ball. Draw a tape diagram to estimate the weight of 3 houseplants.

4. Jane and her 8 friends go apple picking. They share what they pick equally. The total weight of the apples they pick is shown to the right.

   a. About how many kilograms of apples will Jane take home?

27 kg

   b. Jane estimates that a pumpkin weighs about as much as her share of the apples. About how much do 7 pumpkins weigh altogether?

Lesson 8:   Solve one-step word problems involving metric weights within 100 and estimate to reason about solutions.

© 2018 Great Minds®. eureka-math.org

Name _____   Date _____

The weights of a backpack and suitcase are shown below.

 7 kg    21 kg

a. How much heavier is the suitcase than the backpack?

b. What is the total weight of 4 identical backpacks?

c. How many backpacks weigh the same as one suitcase?

Lesson 8:   Solve one-step word problems involving metric weights within 100 and estimate to reason about solutions.

185

© 2018 Great Minds®. eureka-math.org

Name _____   Date _____

Part 1

a. Predict whether each container holds less than, more than, or about the same as 1 liter.

Container 1 holds   (less than) / more than / about the same as   1 liter.   Actual:

Container 2 holds   (less than) / more than / about the same as   1 liter.   Actual:

Container 3 holds   less than / (more than) / about the same as   1 liter.   Actual:

Container 4 holds   less than / more than / (about the same as)   1 liter.   Actual:

b. After measuring, what surprised you? Why?

Part 2

c. Illustrate and describe the process of decomposing 1 liter of water into 10 smaller units.

Lesson 9:   Decompose a liter to reason about the size of 1 liter, 100 milliliters,
10 milliliters, and 1 milliliter.

187

© 2018 Great Minds®. eureka-math.org

d. Illustrate and describe the process of decomposing Cup K into 10 smaller units.

e. Illustrate and describe the process of decomposing Cup L into 10 smaller units.

f. What is the same about decomposing 1 liter into milliliters and decomposing 1 kilogram into grams?

g. One liter of water weighs 1 kilogram. How much does 1 milliliter of water weigh? Explain how you know.

Lesson 9: Decompose a liter to reason about the size of 1 liter, 100 milliliters, 10 milliliters, and 1 milliliter.

Name _____ Date _____

1. Morgan fills a 1-liter jar with water from the pond. She uses a 100-milliliter cup to scoop water out of the pond and pour it into the jar. How many times will Morgan scoop water from the pond to fill the jar?

2. How many groups of 10 milliliters are in 1 liter? Explain.

There are _____ groups of 10 milliliters in 1 liter.

Lesson 9:   Decompose a liter to reason about the size of 1 liter, 100 milliliters,
              10 milliliters, and 1 milliliter.

189

© 2018 Great Minds®. eureka-math.org

Subha drinks 4 large glasses of water each day. How many large glasses of water does she drink in 7 days?

_____

_____

_____

_____

Read   **Draw**   **Write**

Name _____   Date _____

1. Label the vertical number line on the container to the right.
    Answer the questions below.

    a. What did you label as the halfway mark? Why?

    b. Explain how pouring each plastic cup of water helped
        you create a vertical number line.

    c. If you pour out 300 mL of water, how many mL are left
        in the container?

100 mL

2. How much liquid is in each container?

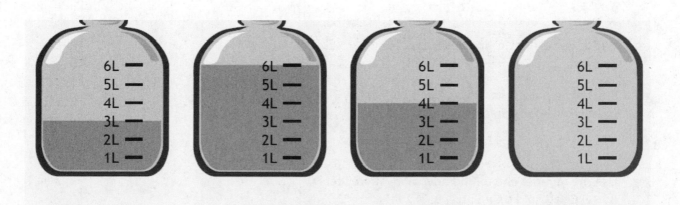

_____   _____   _____   _____

Lesson 10:   Estimate and measure liquid volume in liters and milliliters using the
              vertical number line.

© 2018 Great Minds®. eureka-math.org

193

3. Estimate the amount of liquid in each container to the nearest hundred milliliters.

1000mL	1000mL	1000mL	1000mL
900mL	900mL	900mL	900mL
800mL	800mL	800mL	800mL
700mL	700mL	700mL	700mL
600mL	600mL	600mL	600mL
500mL	500mL	500mL	500mL
400mL	400mL	400mL	400mL
300mL	300mL	300mL	300mL
200mL	200mL	200mL	200mL
100mL	100mL	100mL	100mL

_____ _____ _____ _____

4. The chart below shows the capacity of 4 barrels.

Barrel A	75 liters
Barrel B	68 liters
Barrel C	96 liters
Barrel D	52 liters

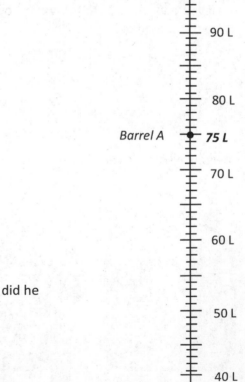

a. Label the number line to show the capacity of each barrel. Barrel A has been done for you.

b. Which barrel has the greatest capacity?

c. Which barrel has the smallest capacity?

d. Ben buys a barrel that holds about 70 liters. Which barrel did he most likely buy? Explain why.

e. Use the number line to find how many more liters Barrel C can hold than Barrel B.

Lesson 10:   Estimate and measure liquid volume in liters and milliliters using the vertical number line.

© 2018 Great Minds®. eureka-math.org

EUREKA MATH

Name _____ Date _____

1. Use the number line to record the capacity of the containers.

Container	Capacity in Liters
A	
B	
C	

Container B ● — 60 L ... 50 L

70 L

60 L

Container B ●

50 L

Container A ●

40 L

2. What is the difference between the capacity of Container A and Container C?

30 L

20 L

Container C ●

10 L

Lesson 10: Estimate and measure liquid volume in liters and milliliters using the
vertical number line.

195

© 2018 Great Minds®. eureka-math.org

Name _____   Date _____

1. The total weight in grams of a can of tomatoes and a jar of baby food is shown to the right.

   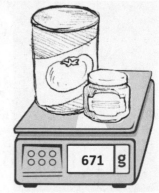

   a. The jar of baby food weighs 113 grams. How much does the can of tomatoes weigh?

   b. How much more does the can of tomatoes weigh than the jar of baby food?

2. The weight of a pen in grams is shown to the right.

   a. What is the total weight of 10 pens?

   b. An empty box weighs 82 grams. What is the total weight of a box of 10 pens?

3. The total weight of an apple, lemon, and banana in grams is shown to the right.

   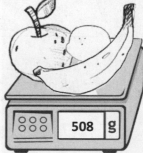

   a. If the apple and lemon together weigh 317 grams, what is the weight of the banana?

   b. If we know the lemon weighs 68 grams less than the banana, how much does the lemon weigh?

   c. What is the weight of the apple?

EUREKA
MATH

Lesson 11:   Solve mixed word problems involving all four operations with grams, kilograms, liters, and milliliters given in the same units.

197

© 2018 Great Minds®. eureka-math.org

4. A frozen turkey weighs about 5 kilograms. The chef orders 45 kilograms of turkey. Use a tape diagram to find about how many frozen turkeys he orders.

5. A recipe requires 300 milliliters of milk. Sara decides to triple the recipe for dinner. How many milliliters of milk does she need to cook dinner?

6. Marian pours a full container of water equally into buckets. Each bucket has a capacity of 4 liters. After filling 3 buckets, she still has 2 liters left in her container. What is the capacity of her container?

Lesson 11: Solve mixed word problems involving all four operations with grams, kilograms, liters, and milliliters given in the same units.

© 2018 Great Minds®. eureka-math.org

Name _____   Date _____

The capacities of three cups are shown below.

Cup A
160 mL

Cup B
280 mL

Cup C
237 mL

a. Find the total capacity of the three cups.

b. Bill drinks exactly half of Cup B. How many milliliters are left in Cup B?

c. Anna drinks 3 cups of tea from Cup A. How much tea does she drink in total?

Lesson 11:   Solve mixed word problems involving all four operations with grams, kilograms, liters, and milliliters given in the same units.

199

© 2018 Great Minds®. eureka-math.org

Name _____ Date _____

1. Work with a partner. Use a ruler or a meter stick to complete the chart below.

Object	Measurement (in cm)	The object measures between (which two tens)...	Length rounded to the nearest 10 cm
Example: My shoe	23 cm	___20___ and ___30___ cm	20 cm
Long side of a desk		_____ and _____ cm	
A new pencil		_____ and _____ cm	
Short side of a piece of paper		_____ and _____ cm	
Long side of a piece of paper		_____ and _____ cm	

2. Work with a partner. Use a digital scale to complete the chart below.

Bag	Measurement (in g)	The bag of rice measures between (which two tens)...	Weight rounded to the nearest 10 g
Example: Bag A	8 g	___0___ and ___10___ g	10 g
Bag B		_____ and _____ g	
Bag C		_____ and _____ g	
Bag D		_____ and _____ g	
Bag E		_____ and _____ g	

Lesson 12:   Round two-digit measurements to the nearest ten on the vertical number line.

201

3. Work with a partner. Use a beaker to complete the chart below.

Container	Measurement (in mL)	The container measures between (which two tens)...		Liquid volume rounded to the nearest 10 mL
Example: Container A	33 mL	___30___ and ___40___ mL		30 mL
Container B		_____ and _____ mL		
Container C		_____ and _____ mL		
Container D		_____ and _____ mL		
Container E		_____ and _____ mL		

4. Work with a partner. Use a clock to complete the chart below.

Activity	Actual time	The activity measures between (which two tens)...		Time rounded to the nearest 10 minutes
Example: Time we started math	10:03	___10:00___ and ___10:10___		10:00
Time I started the Problem Set		_____ and _____		
Time I finished Station 1		_____ and _____		
Time I finished Station 2		_____ and _____		
Time I finished Station 3		_____ and _____		

Lesson 12:   Round two-digit measurements to the nearest ten on the vertical number line.

Name _____ Date _____

The weight of a golf ball is shown below.

a. The golf ball weighs _____.

b. Round the weight of the golf ball to the nearest ten grams. Model your thinking on the number line.

c. The golf ball weighs about _____.

d. Explain how you used the halfway point on the number line to round to the nearest ten grams.

Lesson 12:   Round two-digit measurements to the nearest ten on the vertical
number line.

© 2018 Great Minds®. eureka-math.org

203

The school ballet recital begins at 12:17 p.m. and ends at 12:45 p.m. How many minutes long is the ballet recital?

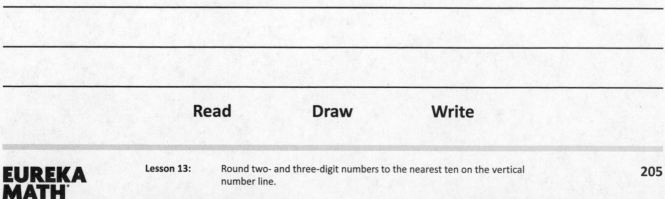

Read   **Draw**   **Write**

Lesson 13:   Round two- and three-digit numbers to the nearest ten on the vertical number line.

205

EUREKA
MATH®

Name _____ Date _____

1. Round to the nearest ten. Use the number line to model your thinking.

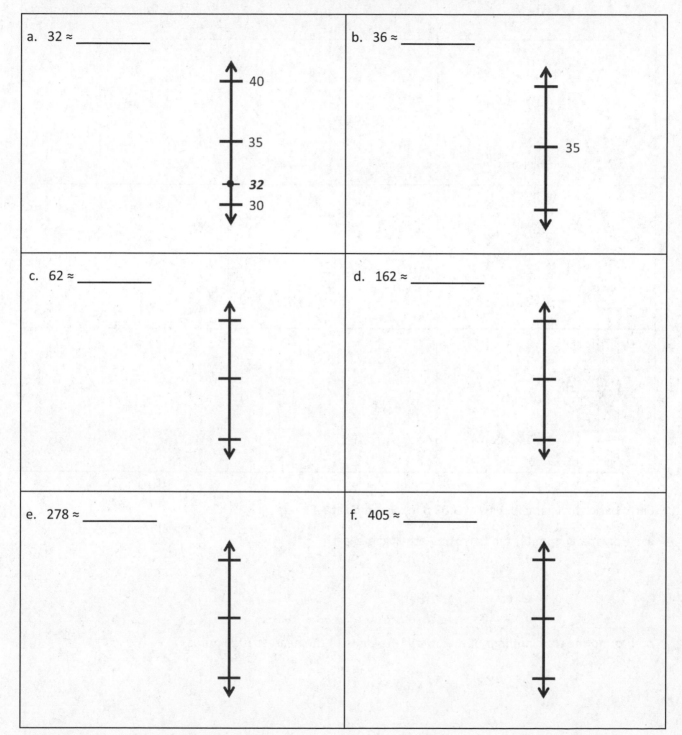

a. 32 ≈ _____

40

35

32

30

b. 36 ≈ _____

35

c. 62 ≈ _____

d. 162 ≈ _____

e. 278 ≈ _____

f. 405 ≈ _____

Lesson 13:   Round two- and three-digit numbers to the nearest ten on the vertical number line.

207

© 2018 Great Minds®. eureka-math.org

2. Round the weight of each item to the nearest 10 grams. Draw number lines to model your thinking.

Item	Number Line	Round to the nearest 10 grams
36 grams		
52 grams		
142 grams		

3. Carl's basketball game begins at 3:03 p.m. and ends at 3:51 p.m.

    a. How many minutes did Carl's basketball game last?

    b. Round the total number of minutes in the game to the nearest 10 minutes.

Lesson 13:   Round two- and three-digit numbers to the nearest ten on the vertical
                   number line.

© 2018 Great Minds®. eureka-math.org

Name _____   Date _____

1. Round to the nearest ten. Use the number line to model your thinking.

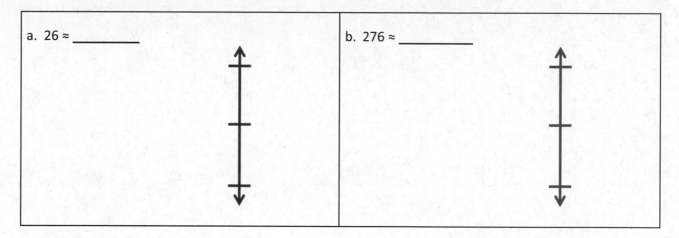

a. 26 ≈ _____

b. 276 ≈ _____

2. Bobby rounds 603 to the nearest ten. He says it is 610. Is he correct? Why or why not? Use a number line and words to explain your answer.

Lesson 13:   Round two- and three-digit numbers to the nearest ten on the vertical
               number line.

© 2018 Great Minds®. eureka-math.org

209

Use place value disks to draw each of the following on a place value chart.

10 tens   10 hundreds   13 tens

13 hundreds   13 tens and 8 ones   13 hundreds 8 tens 7 ones

Read   Draw   Write

EUREKA MATH

Lesson 14:   Round to the nearest hundred on the vertical number line.

211

© 2018 Great Minds®. eureka-math.org

Name _____   Date _____

1.   Round to the nearest hundred. Use the number line to model your thinking.

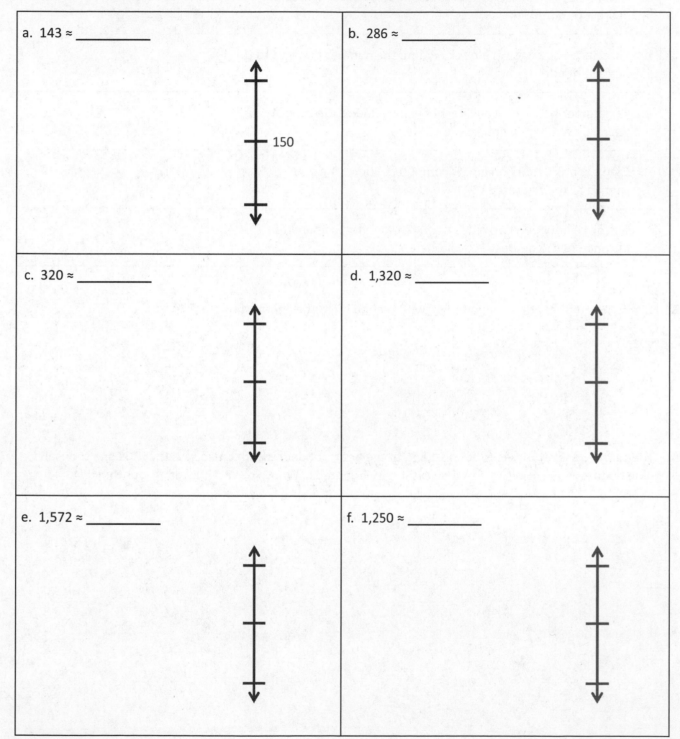

a. 143 ≈ _____

150

b. 286 ≈ _____

c. 320 ≈ _____

d. 1,320 ≈ _____

e. 1,572 ≈ _____

f. 1,250 ≈ _____

EUREKA MATH

Lesson 14:   Round to the nearest hundred on the vertical number line.

213

© 2018 Great Minds®. eureka-math.org

2. Complete the chart.

a. Shauna has 480 stickers. Round the number of stickers to the nearest hundred.	
b. There are 525 pages in a book. Round the number of pages to the nearest hundred.	
c. A container holds 750 milliliters of water. Round the capacity to the nearest 100 milliliters.	
d. Glen spends $1,297 on a new computer. Round the amount Glen spends to the nearest $100.	
e. The drive between two cities is 1,842 kilometers. Round the distance to the nearest 100 kilometers.	

3. Circle the numbers that round to 600 when rounding to the nearest hundred.

527   550   639   681   713   603

4. The teacher asks students to round 1,865 to the nearest hundred. Christian says that it is one thousand, nine hundred. Alexis disagrees and says it is 19 hundreds. Who is correct? Explain your thinking.

Name _____   Date _____

1.   Round to the nearest hundred. Use the number line to model your thinking.

| a. 137 ≈ _____ | b. 1,761 ≈ _____ |

2.   There are 685 people at the basketball game. Draw a vertical number line to round the number of people to the nearest hundred people.

EUREKA
MATH

Lesson 14:   Round to the nearest hundred on the vertical number line.

215

© 2018 Great Minds®. eureka-math.org

unlabeled place value chart

Lesson 14:   Round to the nearest hundred on the vertical number line.

217

© 2018 Great Minds®. eureka-math.org

Use mental math to solve these problems. Record your strategy for solving each problem.

a. 46 mL + 5 mL

b. 39 cm + 8 cm

c. 125 g + 7 g

d. 108 L + 4 L

Read　　　**Draw**　　　**Write**

Lesson 15:　　Add measurements using the standard algorithm to compose larger
　　　　　　　units once.

© 2018 Great Minds®. eureka-math.org

219

Name _____   Date _____

1. Find the sums below. Choose mental math or the algorithm.

    a.   46 mL + 5 mL   b.   46 mL + 25 mL   c.   46 mL + 125 mL

    d.   59 cm + 30 cm   e.   509 cm + 83 cm   f.   597 cm + 30 cm

    g.   29 g + 63 g   h.   345 g + 294 g   i.   480 g + 476 g

    j.   1 L 245 mL + 2 L 412 mL   k.   2 kg 509 g + 3 kg 367 g

 EUREKA MATH®   Lesson 15:   Add measurements using the standard algorithm to compose larger   221
                                                    units once.

© 2018 Great Minds®. eureka-math.org

2. Nadine and Jen buy a small bag of popcorn and a pretzel at the movie theater. The pretzel weighs 63 grams more than the popcorn. What is the weight of the pretzel?

? grams

44 grams

3. In math class, Jason and Andrea find the total liquid volume of water in their beakers. Jason says the total is 782 milliliters, but Andrea says it is 792 milliliters. The amount of water in each beaker can be found in the table to the right. Show whose calculation is correct. Explain the mistake of the other student.

Student	Liquid Volume
Jason	475 mL
Andrea	317 mL

4. It takes Greg 15 minutes to mow the front lawn. It takes him 17 more minutes to mow the back lawn than the front lawn. What is the total amount of time Greg spends mowing the lawns?

Lesson 15:   Add measurements using the standard algorithm to compose larger units once.

EUREKA MATH

Name _____   Date _____

1. Find the sums below. Choose mental math or the algorithm.

    a. 24 cm + 36 cm   b. 562 m + 180 m   c. 345 km + 239 km

2. Brianna jogs 15 minutes more on Sunday than Saturday. She jogged 26 minutes on Saturday.

    a. How many minutes does she jog on Sunday?

    b. How many minutes does she jog in total?

Lesson 15:   Add measurements using the standard algorithm to compose larger   223
               units once.

© 2018 Great Minds®. eureka-math.org

Josh's apple weighs 93 grams. His pear weighs 152 grams. What is the total weight of the apple and the pear?

_____

_____

_____

_____

Read   **Draw**   **Write**

Lesson 16:   Add measurements using the standard algorithm to compose larger units twice.

225

© 2018 Great Minds®. eureka-math.org

Name _____ Date _____

1. Find the sums below.

    a.   52 mL + 68 mL

    b.   352 mL + 68 mL

    c.   352 mL + 468 mL

    d.   56 cm + 94 cm

    e.   506 cm + 94 cm

    f.   506 cm + 394 cm

    g.   697 g + 138 g

    h.   345 g + 597 g

    i.   486 g + 497 g

    j.   3 L 251 mL + 1 L 549 mL

    k.   4 kg 384 g + 2 kg 467 g

Lesson 16:   Add measurements using the standard algorithm to compose larger units twice.

227

2. Lane makes sauerkraut. He weighs the amounts of cabbage and salt he uses. Draw and label a tape diagram to find the total weight of the cabbage and salt Lane uses.

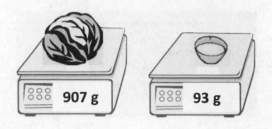

3. Sue bakes mini-muffins for the school bake sale. After wrapping 86 muffins, she still has 58 muffins left cooling on the table. How many muffins did she bake altogether?

4. The milk carton to the right holds 183 milliliters more liquid than the juice box. What is the total capacity of the juice box and milk carton?

Juice Box
279 mL

Milk Carton
? mL

Lesson 16:   Add measurements using the standard algorithm to compose larger units twice.

© 2018 Great Minds®. eureka-math.org

Name _____   Date _____

1. Find the sums.

    a.   78 g + 29 g   b.   328 kg + 289 kg   c.   509 L + 293 L

2. The third-grade class sells lemonade to raise funds. After selling 58 liters of lemonade in 1 week, they still have 46 liters of lemonade left. How many liters of lemonade did they have at the beginning?

The doctor prescribed 175 milliliters of medicine on Monday and 256 milliliters of medicine on Tuesday.

    a. Estimate how much medicine he prescribed in both days.

    b. Precisely how much medicine did he prescribe in both days?

Read   **Draw**   **Write**

Name _____   Date _____

1. a. Find the actual sum either on paper or using mental math. Round each addend to the nearest hundred, and find the estimated sums.

A	B	C
451 + 253 = _____ ___ + ___ = _____ 451 + 249 = _____ ___ + ___ = _____ 448 + 249 = _____ ___ + ___ = _____ Circle the estimated sum that is the closest to its real sum.	356 + 161 = _____ ___ + ___ = _____ 356 + 148 = _____ ___ + ___ = _____ 347 + 149 = _____ ___ + ___ = _____ Circle the estimated sum that is the closest to its real sum.	652 + 158 = _____ ___ + ___ = _____ 647 + 158 = _____ ___ + ___ = _____ 647 + 146 = _____ ___ + ___ = _____ Circle the estimated sum that is the closest to its real sum.

b. Look at the sums that gave the most precise estimates. Explain below what they have in common. You might use a number line to support your explanation.

EUREKA MATH

Lesson 17:   Estimate sums by rounding and apply to solve measurement word problems.

233

© 2018 Great Minds®. eureka-math.org

2. Janet watched a movie that is 94 minutes long on Friday night. She watched a movie that is 151 minutes long on Saturday night.

    a. Decide how to round the minutes. Then, estimate the total minutes Janet watched movies on Friday and Saturday.

    b. How much time did Janet actually spend watching movies?

    c. Explain whether or not your estimated sum is close to the actual sum. Round in a different way, and see which estimate is closer.

3. Sadie, a bear at the zoo, weighs 182 kilograms. Her cub weighs 74 kilograms.

    a. Estimate the total weight of Sadie and her cub using whatever method you think best.

    b. What is the actual weight of Sadie and her cub? Model the problem with a tape diagram.

Lesson 17:   Estimate sums by rounding and apply to solve measurement word problems.

Name _____   Date _____

Jesse practices the trumpet for a total of 165 minutes during the first week of school. He practices for 245 minutes during the second week.

   a.   Estimate the total amount of time Jesse practices by rounding to the nearest 10 minutes.

   b.   Estimate the total amount of time Jesse practices by rounding to the nearest 100 minutes.

   c.   Explain why the estimates are so close to each other.

Lesson 17:   Estimate sums by rounding and apply to solve measurement word
                problems.

© 2018 Great Minds®. eureka-math.org   235

Tara brings 2 bottles of water on her hike. The first bottle has 471 milliliters of water, and the second bottle has 354 milliliters of water. How many milliliters of water does Tara bring on her hike?

_____

_____

_____

Read   **Draw**   **Write**

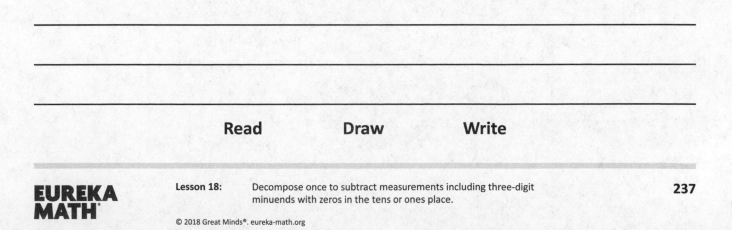

Lesson 18:   Decompose once to subtract measurements including three-digit
              minuends with zeros in the tens or ones place.

237

© 2018 Great Minds®. eureka-math.org

Name _____   Date _____

1. Solve the subtraction problems below.

    a. 60 mL – 24 mL

    b. 360 mL – 24 mL

    c. 360 mL – 224 mL

    d. 518 cm – 21 cm

    e. 629 cm – 268 cm

    f. 938 cm – 440 cm

    g. 307 g – 130 g

    h. 307 g – 234 g

    i. 807 g – 732 g

    j. 2 km 770 m – 1 km 455 m

    k. 3 kg 924 g – 1 kg 893 g

Lesson 18:   Decompose once to subtract measurements including three-digit
             minuends with zeros in the tens or ones place.

239

© 2018 Great Minds®. eureka-math.org

2. The total weight of 3 books is shown to the right. If 2 books weigh 233 grams, how much does the third book weigh? Use a tape diagram to model the problem.

405g

3. The chart to the right shows the lengths of three movies.

The Lost Ship	117 minutes
Magical Forests	145 minutes
Champions	? minutes

a. The movie *Champions* is 22 minutes shorter than *The Lost Ship*. How long is *Champions*?

b. How much longer is *Magical Forests* than *Champions*?

4. The total length of a rope is 208 centimeters. Scott cuts it into 3 pieces. The first piece is 80 centimeters long. The second piece is 94 centimeters long. How long is the third piece of rope?

Lesson 18: Decompose once to subtract measurements including three-digit minuends with zeros in the tens or ones place.

Name _____   Date _____

1. Solve the subtraction problems below.

   a. 381 mL – 146 mL   b. 730 m – 426 m   c. 509 kg – 384 kg

2. The total length of a banner is 408 centimeters. Carly paints it in 3 sections. The first 2 sections she paints are 187 centimeters long altogether. How long is the third section?

Lesson 18:   Decompose once to subtract measurements including three-digit
minuends with zeros in the tens or ones place.

241

© 2018 Great Minds®. eureka-math.org

Jolene brings an apple and an orange with her to school. The weight of both pieces of fruit toegether is 417 grams. The apple weighs 223 grams. What is the weight of Jolene's orange?

_____

_____

_____

_____

Read   **Draw**   **Write**

EUREKA MATH

Name _____   Date _____

1. Solve the subtraction problems below.

   a.   340 cm – 60 cm

   b.   340 cm – 260 cm

   c.   513 g – 148 g

   d.   641 g – 387 g

   e.   700 mL – 52 mL

   f.   700 mL – 452 mL

   g.   6 km 802 m – 2 km 569 m

   h.   5 L 920 mL – 3 L 869 mL

Lesson 19:   Decompose twice to subtract measurements including three-digit
minuends with zeros in the tens and ones places.

245

© 2018 Great Minds®. eureka-math.org

2. David is driving from Los Angeles to San Francisco. The total distance is 617 kilometers. He has 468 kilometers left to drive. How many kilometers has he driven so far?

3. The piano weighs 289 kilograms more than the piano bench. How much does the bench weigh?

Piano
297 kg

Bench
? kg

4. Tank A holds 165 fewer liters of water than Tank B. Tank B holds 400 liters of water. How much water does Tank A hold?

246   **Lesson 19:**   Decompose twice to subtract measurements including three-digit minuends with zeros in the tens and ones places.

© 2018 Great Minds®. eureka-math.org

Name _____   Date _____

1. Solve the subtraction problems below.

    a.   346 m – 187 m   b.   700 kg – 597 kg

2. The farmer's sheep weighs 647 kilograms less than the farmer's cow. The cow weighs 725 kilograms.
    How much does the sheep weigh?

Lesson 19:   Decompose twice to subtract measurements including three-digit
                   minuends with zeros in the tens and ones places.

247

© 2018 Great Minds®. eureka-math.org

Millie's fish tank holds 403 liters of water. She empties out 185 liters of water to clean the tank. How many liters of water are left in the tank?

a. Estimate how many liters are left in the tank by rounding.

b. Estimate how many liters are left in the tank by rounding in a different way.

Read   **Draw**   **Write**

Lesson 20:   Estimate differences by rounding and apply to solve measurement word problems.

249

c. How many liters of water are actually left in the tank?

d. Is your answer reasonable? Which estimate was closer to the exact answer?

Read   **Draw**   **Write**

    Lesson 20:   Estimate differences by rounding and apply to solve measurement word problems.

EUREKA
MATH®

Name _____   Date _____

1. a. Find the actual differences either on paper or using mental math. Round each total and part to the
        nearest hundred and find the estimated differences.

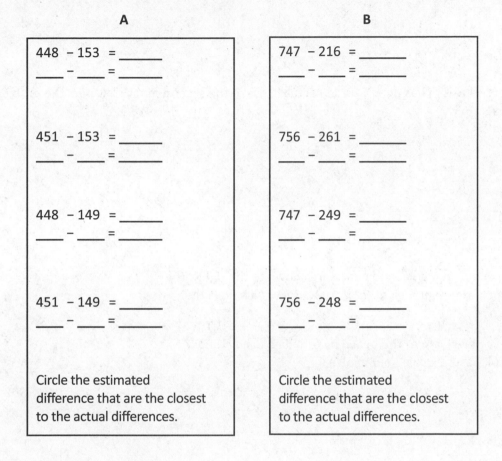

A

448 – 153 = _____
____ – ____ = _____

451 – 153 = _____
____ – ____ = _____

448 – 149 = _____
____ – ____ = _____

451 – 149 = _____
____ – ____ = _____

Circle the estimated
difference that are the closest
to the actual differences.

B

747 – 216 = _____
____ – ____ = _____

756 – 261 = _____
____ – ____ = _____

747 – 249 = _____
____ – ____ = _____

756 – 248 = _____
____ – ____ = _____

Circle the estimated
difference that are the closest
to the actual differences.

b. Look at the differences that gave the most precise estimates. Explain below what they have in
    common. You might use a number line to support your explanation.

Lesson 20:   Estimate differences by rounding and apply to solve measurement
              word problems.

© 2018 Great Minds®. eureka-math.org

251

2. Camden uses a total of 372 liters of gas in two months. He uses 184 liters of gas in the first month. How many liters of gas does he use in the second month?

   a. Estimate the amount of gas Camden uses in the second month by rounding each number as you think best.

   b. How many liters of gas does Camden actually use in the second month? Model the problem with a tape diagram.

3. The weight of a pear, apple, and peach are shown to the right. The pear and apple together weigh 372 grams. How much does the peach weigh?

   a. Estimate the weight of the peach by rounding each number as you think best. Explain your choice.

   b. How much does the peach actually weigh? Model the problem with a tape diagram.

Lesson 20:   Estimate differences by rounding and apply to solve measurement word problems.

© 2018 Great Minds®. eureka-math.org

Name _____   Date _____

Kathy buys a total of 416 grams of frozen yogurt for herself and a friend. She buys 1 large cup and 1 small cup.

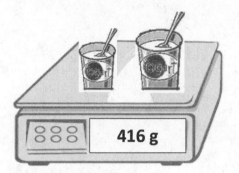

Large Cup	363 grams
Small Cup	? grams

a.   Estimate how many grams are in the small cup of yogurt by rounding.

b.   Estimate how many grams are in the small cup of yogurt by rounding in a different way.

c.   How many grams are actually in the small cup of yogurt?

d.   Is your answer reasonable? Which estimate was closer to the exact weight? Explain why.

Lesson 20:   Estimate differences by rounding and apply to solve measurement   253
               word problems.

© 2018 Great Minds®. eureka-math.org

Gloria fills water balloons with 238 mL of water. How many milliliters of water are in two water balloons? Estimate to the nearest 10 mL and 100 mL. Which gives a closer estimate?

Read   **Draw**   **Write**

EUREKA MATH

Lesson 21:   Estimate sums and differences of measurements by rounding, and then solve mixed word problems.

255

© 2018 Great Minds®. eureka-math.org

Name _____ Date _____

1. Weigh the bags of beans and rice on the scale. Then, write the weight on the scales below.

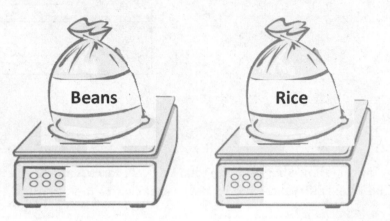

a. Estimate, and then find the total weight of the beans and rice.

Estimate: _____ + _____ ≈ _____ + _____ = _____

Actual: _____ + _____ = _____

b. Estimate, and then find the difference between the weight of the beans and rice.

Estimate: _____ – _____ ≈ _____ – _____ = _____

Actual: _____ – _____ = _____

c. Are your answers reasonable? Explain why.

Lesson 21:   Estimate sums and differences of measurements by rounding, and
then solve mixed word problems.

257

© 2018 Great Minds®. eureka-math.org

2. Measure the lengths of the three pieces of yarn.

   a. Estimate the total length of Yarn A and Yarn C. Then, find the actual total length.

Yarm A	_____ cm ≈ _____ cm
Yarm B	_____ cm ≈ _____ cm
Yarm C	_____ cm ≈ _____ cm

   b. Subtract to estimate the difference between the total length of Yarns A and C, and the length of Yarn B. Then, find the actual difference. Model the problem with a tape diagram.

3. Plot the amount of liquid in the three containers on the number lines below. Then, round to the nearest 10 milliliters.

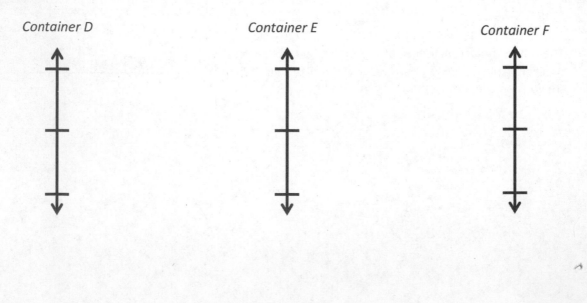

Container D   Container E   Container F

Lesson 21:   Estimate sums and differences of measurements by rounding, and then solve mixed word problems.

a. Estimate the total amount of liquid in three containers. Then, find the actual amount.

b. Estimate to find the difference between the amount of water in Containers D and E. Then, find the
    actual difference. Model the problem with a tape diagram.

4. Shane watches a movie in the theater that is 115 minutes long,
    including the trailers. The chart to the right shows the length in
    minutes of each trailer.

    a. Find the total number of minutes for all 5 trailers.

    b. Estimate to find the length of the movie without trailers.
        Then, find the actual length of the movie by calculating the
        difference between 115 minutes and the total minutes of
        trailers.

Trailer	Length in minutes
1	5 minutes
2	4 minutes
3	3 minutes
4	5 minutes
5	4 minutes
Total	

    c. Is your answer reasonable? Explain why.

EUREKA
MATH

Lesson 21:   Estimate sums and differences of measurements by rounding, and
             then solve mixed word problems.

© 2018 Great Minds®. eureka-math.org

259

Name _____ Date _____

Rogelio drinks water at every meal. At breakfast, he drinks 237 milliliters. At lunch, he drinks 300 milliliters. At dinner, he drinks 177 milliliters.

a. Estimate the total amount of water Rogelio drinks. Then, find the actual amount of water he drinks at all three meals.

b. Estimate how much more water Rogelio drinks at lunch than at dinner. Then, find how much more water Rogelio actually drinks at lunch than at dinner.

EUREKA MATH

Lesson 21:   Estimate sums and differences of measurements by rounding, and then solve mixed word problems.

261

© 2018 Great Minds®. eureka-math.org

Credits

Great Minds® has made every effort to obtain permission for the reprinting of all copyrighted material. If any owner of copyrighted material is not acknowledged herein, please contact Great Minds for proper acknowledgment in all future editions and reprints of this module.